Engineering Risk
and
Hazard Assessment
Volume II

Editors

Abraham Kandel, Ph.D.

Chairman and Professor
Department of Computer Science
Florida State University
Tallahassee, Florida

Eitan Avni, Ph.D.

Research Scientist
Research and Development Division
Union Camp
Princeton, New Jersey

CRC Press, Inc.
Boca Raton, Florida

Library of Congress Cataloging-in-Publication Data

Engineering risk and hazard assessment.

  Includes bibliographies and indexes.
  1. Technology--Risk assessment. I. Kandel, Abraham.
II. Avni, Eitan.
T174.5.E52   1988   363.1   87-20863
ISBN 0-8493-4655-X (set)
ISBN 0-8493-4656-8 (v. 1)
ISBN 0-8493-4657-6 (v. 1)

Direct all inquiries to CRC Press, Inc., 2000 Corporate Blvd., N.W., Boca Raton, Florida, 33431.

© 1988 by CRC Press, Inc.

International Standard Book Number 0-8493-4655-X (Set)
International Standard Book Number 0-8493-4656-8 (Volume I)
International Standard Book Number 0-8493-4657-6 (Volume II)

Library of Congress Card Number 87-20863
Printed in the United States

PREFACE

These volumes deal with the newly emerging field of "Risk and Hazard Assessment" and its application to science and engineering.

The past decade has seen rapid growth in this field but also some "real" disasters in both the U.S. and Soviet Union. It has been the recurrent nightmare of the 20th century: a nuclear power plant explodes, the core begins to melt, a conflagration ignites that spreads a radio-active cloud over the earth.

A malfunction of the coolant system in the core of the Soviet Union's Chernobyl reactor No. 4 may have triggered the violent chemical explosion in the Ukraine.

The Chernobyl disaster has inevitably renewed the debate over the safety of nuclear power plants far beyond Soviet borders. The worst U.S. disaster took place in 1979 at General Public Utility's Three Mile Island (TMI) plant near Harrisburg, Pa. All told, plants in 14 countries have recorded 151 "significant" incidents since 1971, according to a report by the General Accounting Office.

Scientists all over the world point out that there is a limit to how much safety technology can guarantee. Most accidents, including the one at TMI, involved a combination of equipment failure and human error. "In a population of 100 reactors operating over a period of 20 years, the crude cumulative probablility of a [severe] accident would be 45 percent," concluded a recent risk-assessment study by the Nuclear Regulatory Commission, which polices commerical reactors. John G. Kemeny, who headed President Carter's Commission on the TMI accident says: "Something unexpected can always happen. That's the lesson from TMI. All you can do is cut down on the probabilities." As many as half of the safety procedures now routinely conducted in the industry were born after 1979 in the wake of TMI. The industry itself has set up a self-policing body, the Institute of Nuclear Power Operations (INPO). For 15 years hazard analysis has been used in the chemical industry for comparing the risks to employees from various acute hazards. In this edited volume we try to take a better look at hazard assessment and risk analysis in order to improve our understanding of the subject matter and its applicability to science and engineering.

These volumes deal with issues such as short- and long-term hazards, setting priorities in safety, fault analysis for process plants, hazard identification and safety assessment of human-robot systems, plant fault diagnosis expert systems, knowledge based diagnostic systems, fault tree analysis, modeling of computer security systems for risk and reliability analysis, risk analysis of fatigue failure, fault evaluation of complex system, probabilistic risk analysis, and expert systems for fault detection.

It is our hope that this volume will provide the reader not only with valuable conceptual and technical information but also with a better view of the field, its problems, accomplishments, and future potentials.

Abraham Kandel
Eitan Avni
May 1986

THE EDITORS

Abraham Kandel is professor and Chairman of the Computer Science Department at Florida State University in Tallahassee, Florida. He is also the Director of The Institute for Expert Systems and Robotics at FSU. He received his Ph.D. in Electrical Engineering and Computer Science from the University of New Mexico, his M.S. in Electrical Engineering from the University of California, and his B.Sc. in Electrical Engineering from the Technion - Israel Institute of Technology. Dr. Kandel is a senior member of the Institute of Electrical and Electronics Engineering and a member of NAFIPS, the Pattern Recognition Society, and the Association for Computing Machinery, as well as an advisory editor to the international journals *Fuzzy Sets and Systems, Information Sciences,* and *Expert Systems.* He is also the co-author of *Fuzzy Switching and Automata: Theory and Applications* (1979), the author of *Fuzzy Techniques* in *Pattern Recognition* (1982), and co-author of *Discrete Mathematics for Computer Scientists* (1983), *Fuzzy Relational Databases — A Key to Expert Systems* (1984), *Approximate Reasoning in Expert Systems* (1985), and *Mathematical Techniques with Applications* (1986). He has written more than 150 research papers for numerous national and international professional publications in Computer Science.

Eitan Avni is a research scientist at Union Camp, Research and Development Division, Princeton, N.J. and was formerly an assistant professor of Chemical Engineering at Florida State University in Tallahassee, Fla. He received his Ph.D. in chemical engineering from the University of Connecticut. His current research interests include the application of fuzzy sets and artificial intelligence in chemical engineering, and risk and hazard assesment.

CONTRIBUTORS, VOLUME I

Eitan Avni, Ph.D.
Research Scientist
Union Camp
Princeton, New Jersey

L. T. Fan, Ph.D.
Professor and Head
Chemical Engineering Department
Kansas State University
Manhattan, Kansas

Ulrich Hauptmanns, Dr.-Ing.
Hauptprojektleiter
Projektbetreung
Gesellschaft für Reaktorsicherheit
Cologne, West Germany

Koichi Inoue, Dr. of Engineering
Professor
Department of Aeronautical Engineering
Kyoto University
Kyoto, Japan

Abraham Kandel, Ph.D.
Chairman and Professor
Department of Computer Science
Florida State University
Tallahassee, Florida

Trevor A. Kletz, D.Sc.
Professor
Department of Chemical Engineering
Loughborough University of Technology
Leicestershire, England

Hiromitsu Kumamoto, Dr. of Engineering
Research Associate
Department of Precision Mechanics
Kyoto University
Kyoto, Japan

F. S. Lai, Ph.D.
Research Leader
Engineering Section
Grain Marketing Research Laboratory
United States Department of Agriculture
Manhattan, Kansas

Yoshinobu Sato, Master of Engineering
Senior Researcher
Research Institute for Industrial Safety
Ministry of Labour
Tokyo, Japan

Sujeet Shenoi, M.S.
Graduate Research Assistant
Department of Chemical Engineering
Kansas State University
Manhattan, Kansas

CONTRIBUTORS, VOLUME II

Pedro Albrecht
Department of Civil Engineering
University of Maryland
College Park, Maryland

Eitan Avni
Research Consultant
Union Camp
Princeton, New Jersey

Wilker S. Bruce
Department of Computer Science
Florida State University
Tallahassee, Florida

Abraham Kandel, Ph.D.
Chairman and Professor
Department of Computer Science
Florida State University
Tallahassee, Florida

L. F. Pau, D.Sc.
Technical University of Denmark
Lyngby, Denmark

Ahmad Shafaghi, Ph.D.
Senior Engineer
Technica Inc.
Columbus, Ohio

W. E. Vesely, Ph.D.
Senior Staff Scientist
Science Applications International
Columbus, Ohio

Ronald R. Yager
Machine Intelligence Institute
Iona College
New Rochelle, New York

Nur Yazdani, Ph.D.
Assistant Professor
Department of Civil Engineering
College of Engineering
Florida A & M University
and
Florida State University
Tallahassee, Florida

Javier Yllera, Ph.D.
Consultant
Institute of Nuclear Engineering
Technical University
Berlin, West Germany

TABLE OF CONTENTS, VOLUME I

TABLE OF CONTENTS, VOLUME II

Chapter 1

MEASURES OF ACCOUNTABILITY AND CREDIBILITY IN KNOWLEDGE BASED DIAGNOSIS SYSTEMS

Ronald R. Yager

TABLE OF CONTENTS

I. INTRODUCTION

Many applications of knowledge based systems involve the determination of some underlying incidents based upon the observations of some set of symptoms. The classical example of such a system is a medical diagnosis system. Assume E is our set of symptoms and A is a hypothesized set of causes. To determine how good A is as a diagnosis for the sympton set E, we must measure the degree A accounts for the symptoms manifested in E. We call this a measure of *accountability* of A for E. However, this measure of accountability in and of itself is not a good indication of the quality of the diagnosis set A. For it is obvious that one can always increase the size of A to improve the accountability but without forming a better diagnosis. What is needed is an additional measure which we shall call the *credibility* of A given E. This measure of credibility is an indicator of the degree to which the symptoms which must be apparent if the diagnosis is A are contained in the set E. Thus the combination, in particular the conjunction, of these two measures provide an indication of the degree to which A is a good diagnosis in the light of the known evidence E.

In this paper we are primarily concerned with the development of formulations for these two measures. We note that the formulations developed allow for the possibility of multiple causes.

II. MEASURES OF ACCOUNTABILITY

In this section we introduce a measure of accountability which can be used to indicate the degree to which a hypothesized diagnosis can possibly account for a set of observed symptoms.

Let

$$X = \{x_1, x_2, \ldots x_p\}$$

be a set of underlying incidents or causes from among which we have to decide which have occurred or which are the causes of the symptoms or evidence observed. Let

$$Y = \{y_1, y_2, \ldots y_n\}$$

be our set of potential manifestations or symptoms. Essentially our evidence or symptom set will be a subset of Y. Let R be the fuzzy relation over $X \times Y$ in which $R(x,y)$ indicates the maximum possible degree to which incident x can account or explain the occurrence of manifestation y. It is the possibility of y given x. Our observation consists of the fuzzy subset E of Y called the *evidence set*. The evidence set corresponds to the manifestations we observed about the unknown incident which has occurred. For the set E, for any $y \in Y$, $E(y)$ indicates the degree to which the manifestation y is present. The problem we are faced with is that of using the information in the evidence set to help us determine which of the elements in X have occurred. An approach to this problem is to hypothesize the occurrence of some set A of X and determine the degree to which, based upon R, this set can account for the occurrence of the evidence set E.

For any subset A of X we shall denote $S_E(A)$ to indicate the degree to which the occurrence of all the incidents in A can possibly account for all the manifestations found in the evidence set E. We shall call $S_E(A)$ the measure of accountibility of A for the evidence E.

In order for the set A to account for all the evidence in E, the following criteria must be satisfied *for all manifestations, if a manifestation has occurred in E then there must exist some incident in A that is responsible for it.* The measure $S_E(A)$ must be a reflection of the truth of this statement or criteria.

Using Zadeh's theory of approximate reasoning and test-score semantics,[1,2] we can translate the above statement into a measurable quantity to give a formulation for $S_E(A)$. One formulation of this statement is

$$S_E(A) = \min_{y \in Y}[\max_{x \in X}[1 \wedge (1 - E(y) + (R(x, y) \wedge A(x)))]$$

More generally this criteria is seen to be

$$\forall_Y \exists_X (\text{if } E(y) \text{ then } R(x, y) \text{ and } A(x))$$

If we denote the implication operator as $\text{Imp}(a,b)$ and the "and" operator $T(a,b)$, then we get

$$\forall_y \exists_x [\text{Imp}(E(x), T(R(x, y), A(x)))].$$

In the first definition for $S_A(E)$ we have chosen

$$\text{Imp}(a, b) = 1 \wedge (1 - a + b)$$

$$\text{and } T(a, b) = a \wedge b$$

$$\text{where } \wedge = \min.$$

However, in the more general formulation T can be any t-norm[3] and $\text{Imp}(a,b)$ any implication. However, it appears prudent to require that $\text{Imp}(a,b) = 1$ for $b \geq a$.

We note that the measure $S_A(E)$ is somewhat related to Bandler's and Kahout's measure[4] of containment; however, the transformation relation R makes it distinct. In what follows we shall restrict ourselves to the first formulation,

$$S_A(E) = \min_{y \in Y}[\max_{x \in X}[1 \wedge ((1 - E(x) + (R(x, y) \wedge A(x)))]]$$

If we denote $1-E(x)$ as $\bar{E}(x)$, we can more succinctly write this as

$$S_A(E) = \min_{y \in Y}[\max_{x \in X}[1 \wedge (\bar{E}(x) + R(x, y) \wedge A(x))]]$$

Note: If $E = \phi$ then $S_E(A) = 1$

Thus any hypothesized set A completely accounts for the null evidence. It should be noted that

$$0 \leq S_E(A) \leq 1,$$

where the bigger $S_E(a)$ the more A accounts for E.

A potential practical problem which may arise in the implementation of this measure concerns itself with the determination of the elements in the set R. Since Y, which is the set of all potential manifestations, may be huge, this then implies that R will be large. This situation would then require the gathering of a substantial amount of information which may be costly and time consuming. However, for the purpose of evaluating $S_E(A)$, there is no need for all the membership grades in the relation R. At they very most, all that is needed are the $R(x,y)$s for only those pairs (x,y) which have both nonzero membership in E and nonzero membership in A. This follows from the following observations. If $E(y) = 0$, then $\overline{E}(y) = 1 - \overline{E}(y) = 1$ then $(1 \wedge (E(y) + R(x,y) \wedge A(x))) = 1$, regardless of the value of $R(x,y)$. In addition, if $A(x) = 0$ then $A(x) \wedge R(x,y) = 0$ regardless of the value of $R(x,y)$, and hence $1 \wedge (\overline{E}(y) + 0) = \overline{E}(y)$. Further simplication of the calculation $R(x,y)$ is possible. For example, if $A(x)$ is low, then for some ys we can just simply say that $R(x,y)$ is greater than $A(x)$, and hence this implies that $A(x) \wedge R(x,y) = A(x)$, thus freeing us for the requirement of obtaining an exact value for $R(x,y)$. In addition, if for some y there exists one x such that $A(x) \wedge R(x,y) \geq E(y)$, then we don't need to find $R(x,y)$ for any other of the xs. This particular observation is useful when $E(y)$ is low, just knowing that $A(x) \wedge R(x,y)$ is high allows us to assume that $(1 - E(y) + A(x) \wedge R(x,y)) \geq 1$.

It should be stressed that $S_E(A)$ measures the degree to which the occurrence of the *whole set A* is accountable for *all the manifestations in E*. That is, A is considered as a multivalued set rather than a possibility distribution. Thus the measure $S_A(E)$ is useful in situations in which we allow for multiple incidents to occur.

In situations in which it is required, *a priori*, that only one incident in the set X can occur, then we must introduce a different measure of accountability. We shall use $S_E^*(A)$ to indicate the degree to which only *one element* in A can be responsible for *all* the manifestations in the evidence set E. Thus $S_E^*(A)$ will be referred to as the measure of strict (or unique) accountibility of A for the evidence set E. Thus $S_E^*(A)$ can be defined as being required to satisfy the following linquistic specification, "there exists one element in A such that for all the evidence in E this element can be responsible for it."

Again using the translation rules of approximate reasoning and test score semantics we get

$$S_E^*(A) = \underset{x \in X}{Max}[\underset{y \in Y}{Min}[1 \wedge (E^-(y) + (R(x, y) \wedge A(x)))]]$$

As in the case of $S_E(A)$ a more general formulation is possible, that being

$$\underset{x \in X}{\exists} [\underset{y \in Y}{\forall} [Imp(E(y), T(A(x), R(x, y)))]]$$

Theorem: For any sets E and A, and any consistent implementation of $Imp(a,b)$ and $T(a,b)$

$$S_E^*(A) \leq S_E(A)$$

Proof: With

$$S_E^*(A) = \underset{x \in X}{Max}[\underset{y \in Y}{Min}[Imp(E(y), T(A(x), R(x, y)))]]$$

and

$$S_E(A) = \underset{y \in Y}{Min}[\underset{x \in X}{Max}[Imp(E(y), T(A(x), R(x, y)))]]$$

first consider $S_E^*(A)$.
For any $x \in X$

$$\underset{y \in Y}{Min}[Imp(E(y), T(A(x), R(x, y)))] = f(x).$$

However, for all $x \in X$

$$f(x) \leq \underset{y \in Y}{Min}[\underset{x \in X}{Max}[Imp(E(y), T(A(x), R(x, y)))]] \leq S_E(A)$$

Furthermore, since

$$S_E^*(A) = \underset{x \in X}{Max} \; f(x)$$

and we have just shown that for all $x \; \epsilon \; X$

$$f(x) \leq S_E(A)$$

it follows that

$$S_E^*(A) \leq S_E(A).$$

For the special case when $A = \{x\}$, it is reassuring to see that $S_E^*(A) = S_E(A)$.

Theorem: For all E, T, Imp, and any $A = \{a_i/x_i\}$, that is A has at most one element with membership nonzero,

$$S_E(A) = S_E^*(A) = \underset{y \in Y}{Min}[Imp(E(y), T(A(x_1), R(x_1, y)))].$$

Proof: First we recall that for any t-norm, T:

$$T(0, a) = 0$$

$$T(b, a) \geq T(c, a) \quad if \quad b \geq a$$

Thus this implies that

$$Imp(E(y), T(A(x), R(x,y))) = Imp(E(y)) \qquad \qquad for \quad x \neq x_i$$

$$= Imp(E(y), T(A(x_1), R(x_1, y))) \quad for \quad x = x_i$$

Furthermore, for any implication Imp (a,b) $\geq$ Imp (a,c) if B $\geq$ c. Thus for any $y \; \epsilon \; Y$,

$$\underset{x \in X}{Max}[Imp(E(y), T(A(x), R(x, y)))] = Imp(E(y), T(A(x_i), R(x_i, y))).$$

Therefore

$$S_E(A) = \underset{y \in Y}{Min}[Imp(E(y), T(A(x_1), R(x_i, y)))].$$

To calculate S_E^* we see that

$$\underset{y \in Y}{Min}[Imp(E(y), T(A(x), R(x, y)))] = \underset{y \in Y}{Min}[Imp(E(y), 0)] \quad for \quad x \neq x_i$$

$$= \underset{y \in Y}{Min}[Imp(E(y), T(A(x_i), R(x_i, y)))]] \quad for \quad x = x_i$$

Since

$$\underset{y \in Y}{Min}[Imp(E(y), T(A(x_i), R(x_1, y)))] \geq \underset{y \in y}{Min}[Imp(E(y), 0)]$$

and since

$$S_E^*(A) = \underset{x \in X}{Max}[\underset{y \in Y}{Min}[Imp(E(y), T(A(x), R(x, y)))]]$$

it follows that

$$S_E^*(A) = \underset{y \in Y}{Min}[Imp(E(y), T(A(x_i), R(x_i, y)))]$$

Corollary: For the case when $A = \{1/x_i]$ then

$$S_E^*(A) = S_E(A) = \underset{y \in Y}{Min}[Imp[E(y), R(x_i, y)].$$

This follows from the fact that $T(1,a) = a$.
Theorem:

Assume $A = \overset{n}{\underset{i=1}{\cup}} \{A_i\}$,   where $A_i = \{a_i/x_i\}$,   then $S_E^*(A) = Max_1 S_E^*(A_i).$

Proof: First we note that

$$S_E^*(A) = \underset{i=1\ldots n}{\underset{y \in Y}{Max}}[Min[Imp(E(y), T(A(x_i), R(x_i, y)))]]$$

However, as we have just shown,

$$S_E^*(A_i) = \underset{y \in Y}{Min}[Imp(E(y), T(A(x_i), R(x_i, y)))]$$ and the result follows.

Example:

Assume $E = \{1/y_1, 1/y_2\}$ and

$R = \{.3/(x_1, y_1), 1/(x_1, y_2), 1/(x_2, y_1), .6/(x_2, y_2)\}.$

Calculate $S_E(A)$ and $S_E^*(A)$ for $A = \{1/x_1, .7/x_2\}.$

First we note that $\bar{E} = \{0/y_1, 0/y_2\}$ and

$R \cap A = \{.3/(x_1, y_1), 1/(x_1, y_2), .7/(x_2, y_1), .6/(x_2, y_2)\}.$ Therefore

$M = 1 \wedge (\bar{E} + R \cap A) = R \cap A.$

To calculate $S_E(A)$ we note that when $y = y_1$,

$$\underset{x \in X}{Max}[M(x, y)] = Max[.3, .7] = .7$$

and when $y = y_2$

$$\underset{x \in X}{Max}[M(x, y)] = Max[1, .6] = 1$$

hence $S_E(A) = Min(.7, 1) = .7.$

To calculate $S_E^*(A)$ we note that when $x = x_1$

$$Min[M(x, y)] = Min[.3, 1] = .3$$

and when $x = x_2$

$$Min[M(x, y)] = Min[.7, .6] = .6$$

hence $S_E^*(A) = Max[.3, .6] = .6$

III. GENERALIZATION OF ACCOUNTABILITY

In the formulation of our measures of accountability, we required that all the evidence E be accounted for by either all, in the case of $S_E(A)$, or exactly one, in the case of $S_E^*(A)$, of the elements in A. In this section, we shall first show that these two situations are special cases of a more general formulation of the accountability problem. In addition, in some situations it may be desirable to soften the requirement of accounting for all the evidence in E and allow for the accounting of only some portion of the evidence. Thus as a result of this section, we hope to be able to handle statements like *a few of the elements in A account for most of the evidence in E.*

Again let E be our evidence, $E \subset Y$, let $A \subset X$ be a subset of incidents and let R be our possiblity relationship.

Consider the statement, *there exists a subset of A such that for all y ϵ Y, if y is a symptom then there exists an x such that x is contained in this subset and x accounts for y.*

This statement can be written more succinctly: *there exists a subset B of A, such that for all y ϵ Y, if E(y) then $\exists x$, A(x) $\wedge$ B(x) $\wedge$ R(x,y).* Consider the portion

"if   E(y)   then $\exists x$, A(x) $\wedge$ B(x) $\wedge$ R(x, y)"

this can be written as

$$H(B, y) = 1 \wedge (\overline{E}(y) + \bigvee_{x \in X} (A(x) \wedge B(x) \wedge R(x, y)))$$

$$H(B, y) = 1 \wedge (\overline{E}(y) + \underset{x \in X}{Max}(A(x) \wedge B(x) \wedge R(x, y)))$$

$$H(B, y) = \underset{x \in X}{Max}[\overline{E}(y) + (A(x) \wedge B(x) \wedge R(x, y))]$$

The statement *for all y, if y is in E, then there exists an x in A that is contained in B and x, accountants for y* becomes *for all y ϵ Y, H(B,y)* which is equivalent to

$$G(B) = \underset{y \in Y}{Min}[\underset{x \in X}{Max}[\overline{E}(y) + A(x) \wedge B(x) \wedge R(x, y)].$$

Consider now the total statement, this becomes

$\exists$ B $\subset$ A   such that   G(B)

which is expressible as

$$\underset{B \subset A}{Max}\ G(B).$$

Consider now the situation in which instead of allowing all the subsets B of A we are only interested in a subclass of subsets of A, this subclass of A is defined in terms of a predicate M, such that for each subset B of A, $M(B) \in [0,1]$ indicates the degree to which B satisfies the defining condition. We note that the predicate M really defines a fuzzy subset over the set of fuzzy subsets of A. We further note that M can be defined over the set of all fuzzy subsets of X with the understanding that $M(B) = 0$ if B is not contained in A. Consider now the statement *there exists an M subset of A such that for all $y \in Y$, if y is contained in E, then there exists an x contained in A and B that accounts for y*. This statement can be evaluated by finding for each the degree to which it satisfies M and G and then taking the maximum of these, thus

$$S_A^M(E) = \underset{B \in I^X}{Max}[G(B) \wedge M(B)]$$

We shall denote $S_A^M(E)$ as the degree to which an M subset of A accounts for all of E. Special selections of M lead to $S_A(E)$ and S_A^*.

If we define M as

$$M(B) = 1 \quad B = A$$

$$M(B) = 0 \quad B \neq A$$

we are only considering the subset A itself, then

$$S_E^M(A) = Max[G(B) \wedge M(B) \wedge M(B)] = G(A)$$

$$= \underset{y \in Y}{Min}[\underset{x \in X}{Max}[\bar{E}(y) + A(x) \wedge A(x) \wedge R(x, y)]]$$

$$= S_E(A)$$

Thus $S_E(A)$ is the degree to which exactly A accounts for all the evidence in E.

Before proceeding we make the observations that G(B) is a monotonically nondecreasing function of B thus if $B_1 \subset B_2$ then $G(B_2) \geqslant G(B_1)$

Consider next that M defines only the set of subsets of X which have at most one nonzero element, then

$$M(B) = 1 \quad if \quad B(x) \neq 0 \quad for \ only \ one \ element \ in \ X$$

$$M(B) = 0 \quad else$$

and

$$S_E^M(A) = \underset{B}{Max}[G(B) \wedge M(B)] = \underset{\substack{all\ B \\ with\ one \\ element \\ nonzero}}{Max}\ G(B).$$

From the monotonicity property of G(B) we note that for any $x \in X$, if $D = \{a/x\}$ then $G(D) \leqslant G\{1/x\}$, hence

$$S_E^M(A) = \underset{j=1,2,...p}{Max}\ G(B_{xj}), \quad where \quad B = \{1/x_j\}.$$

Therefore

$$S_E^M(A) = \underset{j}{Max}[\underset{y \in Y}{Min}[\underset{x \in X}{Max}[\overline{E}(y) + A(x) \wedge B_j(x) \wedge R(x, y)]$$

however

$$B_j(x) = 1 \quad \text{if} \quad x = x_j$$
$$= 0 \quad \text{if} \quad x \neq x_j$$

Therefore, for any j,

$$\underset{x \in X}{Max}[\overline{E}(y) + A(x) \wedge B_j(x) \wedge R(x, y)] = \overline{E}(y) + A(x_j) \wedge R(x_j, y)$$

Therefore

$$S_E^M(A) = \underset{j}{Max} \underset{y \in Y}{Min}[\overline{E}(y) + A(x_j) \wedge R(x_j, y)] = S_E^*(A)$$

Thus $S_E^*(A)$ is the degree to which a one element subset of A accounts for all the evidence in E.

Having provided this new general formula for the degree to which M type subsets of A account for all the evidence E,

$$S_E^M(A) = \underset{B \subset A}{Max}[M(B) \wedge G(B)]$$

where $\quad G(B) = \underset{y \in Y}{Min}[\underset{x \in X}{Max}[\overline{E}(y) + (A(x) \wedge B(x) \wedge R(x, y))]]$

We can now look at special types of M subsets.

A particularly interesting class of subsets of A are those defined by linguistic quantifiers.[5-7] An example of this would be the statement, "a *few* elements of A account for all the evidence in E."

Linguistic quantifiers such as few, about half, many can be represented as fuzzy subsets. In particular, a linguistic quantifier of type I can be represented as a fuzzy of the real line and a linguistic quantifier of type II can be represented as a fuzzy subset of the unit interval. If the class of fuzzy subsets of interest is defined in terms of a linguistic quantifier then the predicate M defined over the fuzzy subsets B of A can be obtained as follows

$$M(B) = Q(r)$$

where

$$r = \sum_{i=1}^{P} B(x_1) \quad \text{if} \quad Q \text{ is type I}$$

and

$$r = \sum_{i=1}^{P} B(x_1) \bigg/ \sum_{i=1}^{P} A(x_1) \quad \text{if} \quad Q \text{ is type II}$$

Another special situation of interest would be when X = A then we can evaluate statements like, "a few incidents can account for all of the evidence."

Up to the present we have required that however we select the manifestation set from X it must account for *all* the evidence in the set E. In some cases this may be too strong a requirement, and we may be willing to settle for a less stringent requirement such as "most of the evidence is explained by the selected set." In order to accomplish this objective we must look at the function G(B),

$$G(B) = \underset{y_i \in Y}{Min}[\underset{x \in X}{Max}[\overline{E}(y_i) + (A(x) \wedge B(x) \wedge R(x, y_i))]]$$

$$= \underset{y_i \in Y}{Min}[H_B(y_i)]$$

For a given y_i, $H_B(y_i)$ indicates the degree to which B accounts for the symptom y_i. If we desire a less restrictive condition than accounting for "all" the symptoms we must replace the "all" condition, the minimum over Y, by an appropriate quantifier, such as most, at least half, etc. We shall use some of the results obtained in this study[8] to implement this task.

We recall that $E(y_i)$ indicates the degree to which y_i is a symptom. Let $D \subset Y$ be a crisp subset of Y. In this case d_i equals the membership grade of y_i in D. Let Q_2 be the quantifier indicating the proportion of symptoms we desire to satisfy. Then

$$Q_2\left(\sum_{i=1}^{n} (d_i \wedge E(y_i)) \bigg/ \sum_{i=1}^{n} E(y_i)\right)$$

equals the proportion of the symptoms in the set D. Furthermore

$$\underset{y_i \in Y}{Min}[H_B(y_i) \vee \overline{d}_i]$$

indicates the degree to which the symptoms in set D are accountant for by the incidents in B. Combining these two terms via a conjunction operation gives us the degree to which a given subset D of Y satisfies the condition of considering Q_2 symptoms and satisfying these Q_2 symptoms. Finally, in determining whether B satisfies Q_2 of the symptoms we must look at all the subsets D of Y and select the best one. Therefore, the inclusion of a quantifier Q_2 to indicate the proportion of symptoms satisfied by B results in a G(B) term of the following form:

$$G(B) = \underset{D \in 2^Y}{Max}\left[Q_2\left(\sum_{i=1}^{n} (D(y_i) \wedge E(y_i)) \bigg/ \sum_{i=1}^{n} E(y_i)\right) \wedge (Min_{i1}(H_B(y_i) \vee \overline{D}(y_i)))\right]$$

We note in the special case when Q_2 is "all" this reduces to

$$G(B) = Min_i[H_B(y_i)]$$

which is our original formulation.

IV. MEASURE OF CREDIBILITY

In the previous section, a measure of accountability was introduced to indicate the degree to which a set of incidents can account for the evidence. In this section, another measure will be introduced to indicate the degree to which the evidence set E corresponds to the required manifestations of a hypothesized incident set.

One can first introduce a second relationship between the incident set X and the manifestation set Y. For each pair $x \in X$ and $Y \in X$, let $W(x,y)$ indicate the degree to which the occurrence of the incident x requires the appearance of the manifestation y. If one calls W the necessity relationship, thus $W(x,y)$ indicates the degree of necessity of y given x. Whereas $R(x,y)$ indicates the maximal degree to which the occurrence of x can be seen to be responsible for appearance of manifestation y, $W(x,y)$ indicates the minimum degree to which it is necessary for manifestation y to appear if incident x has occurred. We note that for each $x, y \in X \times Y$, $W(x,y) \leq R(x,y)$.

In particular, if incident x is assumed to have occurred, then it is necessary that the evidence set E contain the set:

$$W_x = \{W(x, y_1)/y_1 , W(x, y_2)/y_2,...\}$$

$$W_x = \bigcup_{i=1}^{n} \{W(x, y_i)/y_i\}$$

More generally, assume A is a fuzzy subset of X where

$$A = \{a_1/x_1, a_2/x_2,...a_n/x_n\},$$

that is
$$A(x_i) = a_i.$$

Under the assumption that *all the incidents* in A have occurred, it is required that the evidence set E must contain the set

$$W_A = \bigcup_{i=1}^{n} (a_i \wedge W_{x_i}$$

where $a_i \wedge W_{xi}$ is the fuzzy subset of Y defined by

$$(a_i \wedge W_{x_i}) (y_j) = a_i \wedge W(x_i, y_j)$$

Hence

$$W_A(y) = \underset{x \in X}{Max}[A(x) \wedge W(x, y)]$$

We note that the $\wedge$ (min) operator could be replaced by any t-norm operator, in particular product. However $\wedge$ gives us the largest set W_A and thus is the strongest condition.

We shall define $T_E(A)$ as the degree to which the evidence set E contains all the manifestations required to occur as a result of the assumption that all the incidents in the set A have occured. We shall call $T_E(A)$ the degree of credibility of A given E.

Since $T_E(A)$ measures the degree to which E is contained in the set W_A we can define $T_E(A)$ using the concept of set difference.

Definition: Assume A and B are two fuzzy subsets of X, then

$$D = A - B \quad \text{where}$$

$$D(x) = 0 \vee (A(x) - B(x))$$

In our case we are interested in the difference between $W_A - E$; hence, if we let $D = W_A - E$ we get $D(y) = 0 \vee (W_A(y) - E(y))$. Therefore, $T_E(A)$ can be obtained as

$$T_E(A) = 1 - \underset{y \in Y}{\text{Max}}[D(y)] = 1 - \underset{y \in Y}{\text{Max}}[0 \vee (W_A(y) - E(y))].$$

Furthermore, we note that this is equivalent to

$$T_E(A) = \underset{y \in Y}{\text{Min}}[1 \wedge (1 - W_A(y) + E(y))].$$

Since

$$W_A(y) = \underset{x \in X}{\text{Max}}[A(x) \wedge W(x, y)]$$

we get

$$T_E(A) = \underset{y \in Y}{\text{Min}}[1 \wedge (1 + E(y) - \underset{x \in X}{\text{Max}}(W(x, y) \wedge A(x)))]$$

Note: If $W_A(y) \leq E(y)$ for all $y \in Y$ then $T_E(A) = 1$

Note: If $A = \phi$ then $T_E(A) = 1$

Note: $0 \leq T_E(A) \leq 1$.

It should be stressed that $T_E(A)$ measures the *credibility* of the occurrence of *all* elements of A in light of the evidence set E. That is, $T_E(A)$ is a measure of the degree to which E contains *all* the manifestations necessary for the occurrence of all incidents in A.

In some situations we may be concerned with the occurrence of only one element in the set A. We shall use $T_E^*(A)$ to indicate the degree to which E contains all the manifestations required for the occurrence of at least one element in the set A. $T_E^*(A)$ will be called the lax credibility measure of A given E.

Let

$$H_x = A(x) \wedge W_x$$

be the set of manifestations required by the occurrence of incident $\{A(x)/x\}$. We recall this is a fuzzy subset of Y, where $H_x(y) = A(x) \wedge W(x,y)$. The set difference between H_x and E is

$$D_x = H_x - E$$

where again

$$D_x(y) = 0 \vee (H_x(y) - E(y)).$$

Therefore, the degree to which E is credible under the occurrence of incidence $\{A(x)/x\}$

$$T_E(x) = 1 - \underset{y \in Y}{\text{Max}}\, D_x(y) = \underset{y \in Y}{\text{Min}}[\overline{D}_x(y)]$$

$$= \underset{y \in Y}{\text{Min}}[1 \wedge (1 + E(y) - (W(x, y) \wedge A(x)))].$$

However, $A(x)$ measures the degree to which x is an incident in A and

$$T_E(x) \wedge A(x)$$

measures the degree to which x is an incident in A and the degree to which E covers all the manifestations required by the occurrence of x. From this it follows that

$$T_E(A) = \underset{x \in X}{\text{Max}}[A(x) \wedge \underset{y \in Y}{\text{Min}}[1 \wedge (1 + E(y) - (A(x) \wedge W(x. y)))]]$$

A more general formulation of this measure is possible. Let Q be a linguistic quantifier corresponding to a proportion such as most, many, etc. We are interested in obtaining a measure of the degree to which all of the manifestations required by Q portion the elements in A are satisfied by E, we shall denote this $T_E(A,Q)$.

Let g be a subset of the set $\{1,2, \ldots p\}$ where p is the total number of incidents. In particular $g(i) = 1$ if $i \in g$. Let B(g) be a subset of A formed as follows,

$$B(g)\,(x_1) = A(x_1) \wedge g(i)$$

Under the assumption of the occurrence of the incident set B(g), let H_g be the required manifestations. Thus for any y

$$H_g(y) = \underset{i = 1, \ldots p}{\text{Max}}\, [A(x_i) \wedge g(i) \wedge W(x_i, y)]$$

The difference between H_g and E is denoted

$$D_g = H_g - E$$

where

$$D_g(y) = 0 \vee (H_g(y) - E(y))$$

therefore, the degree to which E is credible under the occurrence of H_g is

$$1 - \underset{y \in Y}{\text{Max}}\, D_g(y) = \underset{y \in Y}{\text{Min}}[\overline{D}_g(y)]$$

Thus for a particular g, this measures the degree to which E covers all the manifestations required. Finally to obtain $T_E(A,Q)$ we get,

$$T_E(A, Q) = \underset{g}{\text{Max}}\left[Q\left(\frac{\sum_{i=1}^{P} (g(i) \wedge A(x_i))}{\sum_{i=1}^{P} A(x_i)} \right) \wedge \underset{y \in Y}{\text{Min}}[D_g(y)] \right]$$

where

$$\overline{D}_g(y) = [1 \wedge (1 + E(y) - \underset{i=1,\dots p}{Max} (A(x_i) \wedge g(i) \wedge W(x_i, y)))]$$

We note that in the special case when Q is "all", then the above requires for the Q portion not to be zero that $\Sigma\ g(i) \wedge A(x_i) = \Sigma\ A(x_i)$ which requires that $g(i) = 1$ for all i. Therefore, this reduces to

$$T_E(A, \text{``all"}) = \underset{y \in Y}{Min}[1 \wedge (1 + E(y) - \underset{i=1,2,\dots p}{Max} [A(x_i) \wedge W(x_i, y))]]$$

which was our original formulation.

As in the case of the accountability measure, the requirement that all manifestations be necessary for the occurrence of a diagnosis set A be can be softened. The method for doing this involves the replacement of the minimization of y by an appropriate quantifier. In particular, the term

$$M_g = \underset{y \in Y}{Min}[1 \wedge (1 + E(y) - \underset{i=1,\dots p}{Max} (A(x_i) \wedge g(i) \wedge W(x_i, y)))]$$

must be modified. We shall use the notation that

$$H_g(y) = \underset{i=1,\dots p}{Max} [A(x_i) \wedge g(i) \wedge W(x_i, y)]$$

In particular if Q_2 is the quantifier used instead of "all" then M_g gets replaced by

$$\underset{D \in 2^Y}{Max}\left[Q_2 \left(\frac{\sum_{i=1}^{n} (D(y_i) \wedge H_g(y_i))}{\sum_{i=1} H_g(y_i)} \right) \wedge Min_i[(1 \wedge (1 - E(y_i) + H_g(y_i)) \vee \overline{D}(y_i)] \right]$$

V. CONCLUSION

We have introduced measures for the accountability and credibility of a diagnosis set A in the face of an evidence set E in the problem of fault diagnosis. It should be noted that the problem of selecting the best diagnosis involves the problem of finding the set A which has the maximal conjunction of credibility and accountability. The solution of this problem involves a fairly complex mathematical programming problem.

ACKNOWLEDGMENTS

This research was in part supported by grants from the National Science Foundation and the Air Force Office of Scientific Research (AFOSR).

REFERENCES

1. **Zadeh, L. A.,** A theory of approximate reasoning, in *Machine Intelligence,* Vol. 9, Hayes, J., Michie, J., and Mikulich, L. I., Eds., John Wiley & Sons, New York, 1979, 149.
2. **Zadeh, L. A.,** Test score semantics for natural languages and meaning representation via proof, in *Empirical Semantics,* Reiger, B. Ed., Bochum, Brockmeyer, Austria, 1981, 281.

3. **Dubois, D.,** Triangular norms for fuzzy sets, in Proc. 2nd Int. Seminar on Fuzzy Sets, Linz, Austria, 1981, 39.
4. **Bandler, W. and Kohout, L.,** Fuzzy power sets and fuzzy implication operators, *Fuzzy Sets and Systems* 4, 13, 1980.
5. **Zadeh, L. A.,** A Computational approach to fuzzy quantifiers in natural languages, *Comput. Math. Appl.,* 9, 149, 1983.
6. **Yager, R. R.,** Quantified propositions in a linguistic logic, *Int. J. Man Mach. Stud.* 19, 195, 1983.
7. **Yager, R. R.,** Quantifiers in the formulation of multiple objective decision functions, *Inf. Sci.,* 31, 107, 1983.
8. **Yager, R. R.,** General multiple objective decision functions and linguistically quantified statements, *Int. J. Man Mach. Stud.,* 21, 389, 1984.

Chapter 2

THE MODELING OF COMPUTER SECURITY SYSTEMS USING FUZZY SET THEORY

Wilker Shane Bruce, Abraham Kandel, and Eitan Avni

TABLE OF CONTENTS

I. INTRODUCTION

The advent of the computer as a problem solving tool is one of the major advances of twentieth century society. The introduction of the computer into any organization can speed the record keeping operations of the organization, allow the storage of large amounts of data in one location and automate the manufacturing processes of the organization. While all of these computer induced changes have the ability to increase the productivity and the profitability of the organization, they also create security problems in the organization for which comprehensive protection programs have not yet been fully perfected.

In general, two different approaches have been taken in the literature to the computer security problem. The first approach to the problem is a discussion of specific protection mechanisms which can be used to combat specific threats to the system. This type of approach can be found in several volumes.[2,4-8,11,14,16,17] The second approach to the problem is a discussion of how to choose a coherent set of mechanisms once threats have been recognized. Some standard volumes which attempt to solve the problem in this manner are described in References 3, 13, and 15.

The development of a computer security program in any organization must deal with three basic issues. These are (1) what are the specific security problems of the organization; (2) what mechanisms are available which will provide protection for the computer related assets of the organization; and (3) what degree of security is appropriate for each of the assets of the organization. A corollary issue which must be faced is the determination of the degree of protection which any specific mechanism might provide. However, the complexity of modern day computer systems often makes it impossible to completely understand both the full effect of an undesired action upon the system and the actual degree of protection which any specific mechanism might give the system. As a result of the uncertainty caused by the complexity of the system, generalized computer security models have been developed.

It is the nature of the modeling process that some information inherent in the system is lost during modeling. This is not undesirable, since one of the purposes of creating a model is to generalize a system whose complexity makes it incomprehensible. However, the uncertainty which results from the loss of information in the model must be handled by the methods used in the model or the results which are obtained from the model will not be reliable. One method which has been developed for working with systems which have lost information in the modeling process is fuzzy set theory. Since fuzzy sets do not require strict partitioning of objects into groups, it is possible to handle situations where one desires the creation of sets where different elements of the set have differing amounts of membership in the set. An example of this type of set is the set of all mechanisms which will stop a specific undesired action, where the effectiveness of any specific mechanism in stopping the undesired action is not fully understood.

In the past, both probabilistic methods and fuzzy methods have been applied to different aspects of the computer security problem. However, fuzzy methods have not yet been applied to the problem of creating an entire computer security program. Probabilistic methods have been applied to this problem. The result of this application is the *probabilistic risk analysis model*. It will be the purpose of this chapter to create a version of the risk analysis model which uses fuzzy methods. The starting point for the development of the fuzzy model is the probabilistic risk analysis model of Brocks.[1]

II. FUZZY TOOLS

In computer security, as in many other areas of concern to the computer scientist, there is the problem of how to deal with imprecision in the model of the system. This phenomenon occurs in several ways. First, present day systems are often so complex as to resist all efforts at total comprehension. Second, even when complex systems are totally understandable, often the time constraints involved in the real time applications of these systems require that system security be treated in terms of a generalized model of the system as a whole. Given that these reasons for treating systems in an imprecise manner are valid, it would be foolish to attempt to provide security for systems without using methods that help control the inherent risks involved in generalizing the system model. One method which has been developed for dealing with imprecise systems which require a measure of subjectivity in their treatment is fuzzy set theory. It is the purpose of this chapter to define some of the tools which are derived by the application of fuzzy set theory to statistics.[9-10]

The following discussion is based on Zadeh's work.[18-22]

A *fuzzy set* consists of objects and their respective grades of membership in the set. The *grade of membership* of an object in the fuzzy set is given by a subjectively defined *membership function*. The value of the grade of membership of an object can range from 0 to 1 where the value of 1 denotes full membership, and the closer the value is to 0, the weaker is the object's membership in the fuzzy set.

As an example of the use of fuzziness, consider a set of swimmers. If X swims a mile daily whereas Y swims only half a mile three times a week, we can subjectively assign grades of membership in the set of swimmers for X and Y to be 1.0 and 0.3, respectively. It is a natural way of expressing the fact that X is much more of a swimmer than Y.

Definition 1: Let U be the universe of discourse, with the generic element of U denoted by u. A *fuzzy subset* F of U is characterized by a *membership function* $m_F: U \rightarrow [0,1]$, which associates with each element u of U a number $m_F(u)$ representing the *grade of membership* of u in F. F is denoted as $\{(u, m_F(u)) \mid u \; U\}$. In this paper we use interchangeably m_F and χ to represent grades of memberships.

Other widely used notations are

$$F = \int_U m_F(u)/u \quad \text{when U is a continuum} \tag{1}$$

$$F = m_F(u_1)/u_1 + \ldots + m_F(u_n)/u_n$$

$$= \sum_{i=1}^{n} m_F(u_i)/u_1 \tag{2}$$

when U is a finite or countable set of n elements.

Definition 2: The *support* of F is the set of points in U at which $m_F(u)$ is positive.

Definition 3: The *height* of F is the supremum of $m_F(u)$ over U,

$$\text{hgt(F)} = \sup_{u \in U} m_F(u) \tag{3}$$

Definition 4: A fuzzy set of F is said to be *normal* if its height is unity, that is, if

$$\sup_{u \in U} m_F(u) = 1$$

Otherwise F is *subnormal*. (It maybe noted that a subnormal fuzzy set F may be normalized by dividing m_F by hgt(F).)

Definition 5: Let A be a fuzzy subset of U, and let b be another fuzzy or nonfuzzy (ordinary) subset of U. A is a *subset* of B or

$$A \subset B \Leftrightarrow m_A(u) < m_B(u), \ u \in U. \tag{4}$$

III. OPERATIONS ON FUZZY SETS

Throughout the forthcoming discussion, the symbols # and & stand for max and min, respectively; thus, for any real a, b

$$a \ \# \ b = \max (a, b) = a \qquad \text{if } a > b \tag{5}$$

$$= b \qquad \text{if } a < b$$

and

$$a \ \& \ b = \min (a, b) = a \qquad \text{if } a < b \tag{6}$$

$$= b \qquad \text{if } a > b$$

Consistent with this notation, the symbol #z means "supremum over the values of z" and &z should be read as "infimum over the values of z" where z ϵ z.

Among the basic operations that can be performed on fuzzy sets are the following:

Definition 6: The *complement* of a fuzzy set A is denoted by $\bar{A}$ (or by A') and is defined by

$$\bar{A} = \int_v [1 - m_A(u)]/u \tag{7}$$

The operation of complementation is equivalent to negation. Hence, if A is a label for a fuzzy set, then NOT A would be interpreted as $\bar{A}$.

Definition 7: The *union* of fuzzy sets A and B is denoted by A + B (or by A $\cup$ B) and is defined by

$$A + B = \int_U [m_A(u) \ \# \ m_B(u)]/u \tag{8}$$

The union corresponds to the connective OR. Thus, if A and B are labels of fuzzy sets, then A OR B is expressed as A + B.

Definition 8: The *intersection* of fuzzy sets A and B is denoted by A $\cap$ B and is defined by

$$A \cap B = \int_U [m_A(u) \ \& \ m_B(u)]/u \tag{9}$$

The intersection corresponds to the connective AND. Hence A AND B is interpreted as A $\cap$ B.

COMMENT: It should be pointed out that # and & are not the only operations used as interpretations of the union and intersection, respectively. In particular, when AND is identified with & (i.e., min), it represents a "hard" AND in the sense that no trade-off is allowed between its operands. By contrast, an AND that is interpreted in terms of the arithmetic product of the operands, acts as a "soft" AND. Which of these or other possible interpretations is more appropriate depends on the applications in which OR and AND are used.

Definition 9: The *product* of fuzzy sets A and B is denoted by AB and is defined by

$$AB = m_A(u) \cdot m_B(u)/u \tag{10}$$

Thus, A^p, where p is any positive number, is defined by

$$A^p = \int_U [m_A(u)]^p/u \tag{11}$$

Similarly, if w is any nonnegative real number such that $w'hgt(A) \leq 1$, then

$$wA = \int_U w'm^a(u)/u \tag{12}$$

Two operations that are defined as special cases of Equation 11 are useful in the representation of linguistic hedges.
The operation of *concentration* is denoted by CON(A) and is defined by

$$CON(A) = A^2 \tag{13}$$

The concentration is an interpretation of VERY. Thus, if A is a label of a fuzzy set, then VERY A corresponds to CON(A).
The operation of *dilation* is denoted by DIL(A) and is expressed by

$$DIL(A) = A^{0.5} \tag{14}$$

If A is a label of a fuzzy set, then APPROXIMATELY A is interpreted as DIL(A).

Example 1: Given the universe of discourse

$$U = 1 + 2 + \ldots + 8, \tag{15}$$

and fuzzy subsets

$$A = .8/3 + 1/5 + .6/6 \quad \text{and} \quad B = .7/3 + 1/4 + .5/6 \tag{15a}$$

then

$$\tilde{A} = 1/1 + 1/2 + .2/3 + 1/4 + .4/6 + 1/7 + 1/8$$
$$A + B = .8/3 + 1/4 + 1/5 + .6/6$$
$$A \& B = .7/3 + .5/6$$

$$AB = .56/3 + .3/6$$

$$A^3 = .512/3 + 1/5 + .216/6$$

$$.6B = .42/3 + .6/4 + .3/6$$

$$CON(A) = .64/3 + 1/5 + .36/6$$

$$DIL(B) = .84/3 + 1/4 + .71/7 \tag{16}$$

Definition 10: If $A_1, \ldots, A_n$ are fuzzy subsets of $U_1, \ldots, U_n$, respectively, the *Cartesian product* of $A_1, \ldots, A_n$ is denoted by $A_1 \times \ldots \times A_n$ is defined as a fuzzy subset of $U_1, \ldots, U_n$ whose membership function is expressed by

$$m_{A_1} \times \ldots \times m_{A_n}(u_1, \ldots, u_n) = m_{A_1}(u_1) \& \ldots \& m_{A_n}(u_n). \tag{17}$$

Thus, $A_1 \times \ldots \times A_n$ can be written as

$$U_1 \times \ldots \times U_n[m_{A_1}(u_1) \& \ldots \& m_{A_n}(u_n)]/(u_1, \ldots, u_n) \tag{18}$$

The concept of the Cartesian product will be further referenced in the discussion of fuzzy relations.

Example 2: Given $U_1 = U_2 = 1 + 2 + 3$, $A_1 = .5/1 + 1./2 + .6/3$ and $A_2 = 1./1 + .6/2$, then

$$A_1 \times A_2 = .5/(1, 1) + 1./(2, 1) + .6/(3, 1)$$

$$+ .5/(1, 2) + .6/(2, 2) + .6/(3, 2) \tag{19}$$

Definition 11: If $A_1, \ldots, A_n$ are fuzzy subsets of $U_1, \ldots, U_n$ (not necessarily distinct), and $w_1, \ldots, w_n$ are nonnegative weights such that $\sum_{i=1} w_i = 1$, then the convex combination of $A_1, \ldots, A_n$ is a fuzzy set A whose membership function is defined by

$$m_A = w_1 M_{A_1} + \ldots + w_n m_{A_n} \tag{20}$$

where $+$ denotes the arithmetic sum. The concept of a convex combination is useful in the representation of linguistic hedges such as *essentially, typically,* etc., which modify the weights associated with components of a fuzzy set. The weights can also be interpreted as coefficients of importance of the components of a fuzzy set A "built" from fuzzy sets $A_1, \ldots, A_n$.

Definition 12: If A is a fuzzy subset of U, then a *t-level set* of A is a nonfuzzy set denoted by A_t which comprises all elements of U whose grade of membership in A is greater or equal to t. In symbols,

$$A_t = \{u: m_A(u) \geq t\} \tag{21}$$

A fuzzy set A may be decomposed into its level-sets through the *resolution identity:*

$$A = \int_0^1 tA_t \tag{22}$$

or

$$A = \sum_t tA_t \tag{23}$$

where tA_t is the product of a scalar t with the set A_t [in the sense of Equation 12], and $\int_0^1$ (or $\sum_t$) is the union of the A_t sets, with t ranging from 0 to 1.

Since Equations 22 or 23 may be interpreted as representations of a fuzzy set as a union of its constituent fuzzy singletons (m_i/u_i), it follows from the definition of the union (definition 7) that if in the representation of A we have $u_i = u_j$, then we can make the substitution expressed by

$$m_i/u_i + m_j/u_j = (m_i \# m_j)/u_i \tag{24}$$

For example, given

$$A = .4/a + .7/a + .6/b + .3/b,$$

A may be rewritten as

$$A = (.4 \# .7)/a + (.6 \# .3)/b = .7/a + .6/b$$

Or conversely,

$$m_i/u_i = (\# m_j)/u_i; \ 0 \leqslant t \leqslant m_i; \ m_j \in [t, m_i] \tag{25}$$

For example,

$$.4/a = (.1 \# .2 \# .3 \# .4)/a; \ t = .1$$

Thus the resolution identity may be viewed as the result of combining together those terms in Equation 1 or Equation 2 which fall into the same level-set.

More specifically, suppose that A is represented in the form

$$A = .1/a + .3/b + .5/c + .9/d + 1/e.$$

Then by using Equation 25, A can be rewritten as

$$\begin{aligned}
A = \ &.1/a + .1/b + .1/c + .1/d + .1/e \\
&+ .3/b + .3/c + .3/d + .3/e \\
&+ .5/c + .5/d + .5/e \\
&+ .9/d + .9/e \\
&+ 1/e
\end{aligned}$$

or

$$\begin{aligned}
A = \ &.1(1/a + 1/b + 1/c + 1/d + 1/e) + \\
&.3(1/b + 1/c + 1/d + 1/e) +
\end{aligned}$$

$$.5(1/c \; + \; 1/d \; + \; 1/e) \; +$$

$$.9(1/d \; + \; 1/e) \; +$$

$$1(1/e)$$

which is in the form of Equation 23. Using Equation 21, the level-sets are given by

$$A_{.1} = a + b + c + d + e$$

$$A_{.3} = b + c + d + e$$

$$A_{.5} = c + d + e$$

$$A_{.9} = d + e$$

$$A_1 = e.$$

As will be seen later, the resolution identity provides a convenient way of generalizing various concepts associated with nonfuzzy sets to fuzzy sets. The level-sets of a fuzzy set will be used as the basis in establishing the response set to a given query given a threshold of acceptance, or t-level.

Definition 13: The operation of *fuzzification* is used in transforming a nonfuzzy set into a fuzzy set of increasing fuzziness of a fuzzy set. Thus, a *fuzzifier F* applied to a fuzzy subset A of U produces a fuzzy subset F(A;K) which is expressed by

$$F(A;K) = \int_U m_A(u) \; K(u) \tag{26}$$

where the fuzzy set K(u) is the *kernel of F*, i.e., the result of applying F to a singleton 1/u:

$$K(u) = F(1/u;K) \tag{27}$$

Here, $m_A(u)K(u)$ represents the product of a scalar $m_A(u)$ and the fuzzy set K(u) [see Equation 12], and $\int$ is the union of the family of fuzzy sets $m_A(u)K(u)$, $u \in U$. In effect, Equation 26 can be viewed as an integral representation of a linear operator, with K(u) being the counterpart of the impulse response.

The operation of fuzzification finds an important application in the definition of linguistic hedges such as MORE OR LESS, SOMEWHAT, SLIGHTLY, MUCH, etc. For example, if a set A of positive numbers is labeled as POSITIVE, then SLIGHTLY POSITIVE is a label for a fuzzy subset of the real line. In this case, SLIGHTLY is a fuzzifier which transforms POSITIVE into SLIGHTLY POSITIVE. However, the effect of the fuzzifier SLIGHTLY cannot be expressed in the form of Equation 26, as is a case with some other fuzzifiers.

IV. THE EXTENSION PRINCIPLE

The extension principle, introduced by Zadeh,[22] is one of the most important concepts in the fuzzy set theory. Application of this principle transforms any mathematical relation between nonfuzzy elements to deal with fuzzy entities.

Definition 14: *The Extension Principle.* Let $A_1, \ldots, A_n$ be fuzzy sets over $U_1, \ldots, U_n$ respectively, with their Cartesian product defined by Equation 18. Let f be a function from $U_1 \times \ldots \times U_n$ to Y. The fuzzy image B of $A_1, \ldots, A_n$ through f has a membership function:

$$m_B(y) = \left\{ \sup_{(u_1,\ldots,u_n) \in U_1 \times \ldots \times U_n} \min_{i=1,\ldots,n} m_{A_i}(u_i) \right\}, \forall y \in Y \qquad (28)$$

under the constraint $y = f(u_1, \ldots, u_n)$ and the additional condition $m_B(y) = 0$ when $f^{-1}(y) = \{u_1, \ldots, u_n | y = f(u_1, \ldots u_n)\} = 0$

The extension principle can be applied in the composition of fuzzy sets of functions, in the algebra of fuzzy numbers, in defining the fuzzy maximum value of a function over a fuzzy domain, and in multivalued logic and other important areas.

Among more specific applications let us mention the *consistency degree* of two fuzzy sets A and B that can be expressed as

$$\sup_{x=y} \{\min(m_A(x), m_B(y))\} = \text{hgt}(A \cap B) \qquad (29)$$

It is the application of the extension principle to the equality.

V. POSSIBILITY THEORY

In dealing with computer security considerations, it is often the case that the data used are neither exact nor lend themselves to exact analysis. This so called "soft data" can be inexact in several ways. First, it may not be possible to determine whether or not a piece of information which will enable an individual to overcome the security measures of a system is available to that individual. Second, even if it is possible to make an exact verification of whether this information is available, it may not be within the ability of the system to obtain this data within a reasonable cost. Often probability theory has been used to handle soft data in the security structures. However, probability theory has the inherent difficulty that there is often a difference between what is probable and what is possible.

The following example is used to explain the relation between fuzziness and possibility.[19]

Example 3: Consider a non fuzzy statement (or proposition p) $p \hat{=} X$ is an integer in the interval [0, 5]. The interpretation of the proposition p asserts that:

1.   It is possible for any integer in the interval [0, 5] to be a value of X.
2.   It is not possible for any integer outside of the interval [0, 5] to be a value of X.

In other words, q induces a possibility distribution Π_X which associates with each integer $u \in [0, 5]$ the possibility that u could be a value of X. Hence,

$$\Pi_X \hat{=} \begin{cases} \text{Poss } \{X = u\} = 1 & \text{for } 0 \leqslant u \leqslant 5 \\ \\ \text{Poss } \{X = u\} = 0 & \text{for } u < 0 \text{ or } u > 5 \end{cases}$$

where Poss $\{X = u\}$ stands for "The possibility that X may assume the value of u".

Now, "fuzzify" the proposition p:

$$q \stackrel{\triangle}{=} X \text{ is a small integer}$$

where "small integer" is a fuzzy set defined in the universe of positive integers as

$$\text{SMALL INTEGER} = 1/0 + 1/1 + .9/2 + .7/3 + .5/4 + .2/5$$

where 0.7/3 signifies that the grade of membership of the integer 3 in the fuzzy set SMALL INTEGER — or, equivalently, the compatibility of the statement that 3 is a SMALL INTEGER — is 0.7. Consequently, the interpretation of the proposition q can be fuzzified:

It is possible for any integer to be a SMALL INTEGER with the possibility of X taking a value of u being equal to the grade of membership of u in the fuzzy set SMALL INTEGER. In other words, q induces a possibility distribution Π_X which associates with each integer the possibility that u could be a value of X equal to the grade of membership of u in the fuzzy set SMALL INTEGER. Thus,

$$\text{Poss } \{X = 0\} = \text{Poss } \{X = 1\} = 1$$

$$\text{Poss } \{X = 2\} = .9$$

$$\text{Poss } \{X = 3\} = .7$$

$$\text{Poss } \{X = 4\} = .5$$

$$\text{Poss } \{X = 5\} = .2$$

$$\text{Poss } \{X = u\} = 0 \quad \text{for} \quad u < 0 \quad \text{or} \quad u > 5.$$

More generally, the above interpretation of a fuzzy proposition can be stated in the following postulate.

A. Possibility Postulate

Definition 5: *Possibility Postulate.* If X is a variable that takes values in U, and F is a fuzzy subset of U characterized by a membership function m_F, then the proposition

$$q \stackrel{\triangle}{=} X \text{ is F} \tag{30}$$

induces a *possibility distribution* Π_X which is equal to F, i.e.,

$$\Pi_X = F \tag{31}$$

implying that

$$\text{Poss } \{X = u\} = m_F(u) \quad \text{for all} \quad u \in U. \tag{32}$$

In essence, the possibility distribution of X engenders a fuzzy set which serves to define the possibility that X could take any value u in U. It is important to note that since $\Pi_X = F$, the possibility distribution depends on the definition off and hence is purely subjective in nature.

Correspondingly, the *possibility distribution function* associated with X (or the possibility distribution function of Π_X) is denoted by p_X and is defined to be numerically equal to the membership function of F, i.e.,

$$p_X = m_F. \tag{33}$$

Thus, $p_X(u)$, the *possibility* that $X = u$, is postulated to be equal to $m_F(u)$.

Equation 31 is referred to as a *possibility assignment equation* because it signifies that the proposition "X is F" translates into the assignment of a fuzzy set F to the possibility distribution of X, or

$$q \overset{\triangle}{=} X \text{ is } F \rightarrow \Pi_X = F \tag{34}$$

where $\rightarrow$ stands for "translates into".

More generally, the possibility assignment equation corresponding to a proposition of the form

$$\text{"N is F"} \tag{35}$$

where F is a fuzzy subset of a universe of discourse U, and N is the name of (1) a variable, (2) a fuzzy set, (3) a proposition, or (4) an object, and may be expressed as

$$\Pi_{X(N)} = F \tag{36}$$

or, more simply,

$$\Pi_X = F. \tag{36a}$$

X is either N itself (when N is a variable) or a variable that is explicit or implicit in N, with X taking values in U.

Example 4:

$$q \overset{\triangle}{=} \text{Tom is old.}$$

Here $N = \text{Tom}$, $X = \text{Age(Tom)}$, and "old" is a fuzzy set defined on $U = \{u: 0 \leqslant u \leqslant 100\}$ with u signifying Age and characterized by a membership function m_{OLD}. Hence,

$$q \overset{\triangle}{=} \text{Tom is old} \rightarrow \Pi_{\text{Age(Tom)}} = \text{OLD.}$$

It is clear that the concept of a possibility distribution is closely related to fuzzy sets; therefore, possibility distributions can be manipulated by the rules applicable to fuzzy sets.

B. Fuzzy Restriction

The notion of a possibility distribution bears also a close relation to the concept of a *fuzzy restriction*. Hence, the mathematical apparatus for fuzzy restrictions, the calculus of fuzzy restrictions, can be used as a basis for the manipulations of possibility distributions.

First, let us define the concept of a fuzzy restriction.

Definition 16: Let X be a variable which takes values in a universe of discourse U, and let $X = u$ signify that X is assigned the value of u where u is an element of U. Let F be a fuzzy subset of U which is characterized by a membership function m_F. The F is a *fuzzy restriction on X* (or *associated with X*) if F acts as an elastic constraint on the value that may be assigned to X. In other words, the assignment of a value of u to X has the form

$$X = u : m_F(u) \tag{37}$$

Table 1
THE POSSIBILITY AND PROBABILITY
DISTRIBUTION ASSOCIATED WITH X

u	1	2	3	4	5	6	7
Poss{X = u}	1	1	1	0.8	0.4	0.1	0
Prob{X = u}	0.3	0.4	0.2	0.1	0	0	0

where $m_F(u)$ is interpreted as the degree to which the constraint represented by F is satisfied when u is assigned to X, or $m_F(u)$ can be thought of as a degree of compatibility of u with F. It follows that $[1 - m_F(u)]$ is the degree to which the constraint represented by F is not satisfied, or it is the degree to which the constaint must be stretched so that the assignment of u to X is possible.

It is important to note that the fuzzy set per se is not a fuzzy restriction. To be a fuzzy restriction, it must be acting as a constraint on the values of a variable. In other words, a variable X can assume values in U depending on the definition of the fuzzy set F.

VI. POSSIBILITY - PROBABILITY RELATIONSHIP

Both the concepts of possibility and probability are means of representing and manipulating uncertainty or imprecision. A simple example is used to illustrate the difference between probability and possibility.

Example 5: Suppose that a seven-member family owns a four-set car. Now let us consider how many passengers can ride in the car. This corresponds to the statement "X passengers ride in a car" where X takes on values in U = {1, 2, . . . 7}. We can associate a possibility distribution with X by interpreting Poss{x = u} as the degree of ease with which u passengers can ride in a car. Let us also associate a probability distribution with X where Prob{x = u} stand for the probability that u people will ride in the car. The values of Poss{X = u} and Prob{X = u} are assessed subjectively and are shown in the Table 1.

Some intrinsic difference between possibility and probability can be observed from this table. While the probabilities have to sum up to 1 over U, the possibility values are not so restricted. Also notice that the possibility that 3 passengers will ride in the car is 1, while the probability is quite small, i.e., 0.2. Thus, a high possibility does not imply a high probability, nor does a low degree of probability imply a low degree of possibility. However, lessening the possibility of an event tends to lessen its probability, but the converse is not true. Furthermore, if an event is impossible, it is bound to be improbable. This heuristic relationship between possibilities and probabilities is called the *possibility/probability consistency principle.*

Example 6: As another example of the difference between probability and possibility, consider again the statement "Tom is old" with the translation having the form

$$\text{Tom is old} \rightarrow R(\text{Age}(\text{TOM})) = \text{OLD}.$$

Let us assume that the fuzzy set old is subjectively defined on U = {u : 0 ⩽ u < 100}. Hence $\Pi_{A(X)}(u)$ may have sample values given in Table 2.

To associate a probability distribution with A(X), it is necessary to translate the fuzzy concept old into a "hard" (nonfuzzy nonelastic) concept. Let us adopt the definition that any age over or equal to 60 is "old". Then, the probabilities assume the values as shown in Table 3.

Table 2
POSSIBILITY DISTRIBUTION ASSOCIATED WITH *OLD*

u	10	20	25	30	35	40	50	60	...
$\Pi_{A(X)}$ (u)	0	0.2	0.3	0.5	0.8	0.9	1	1	...

Table 3
PROBABILITY DISTRIBUTION ASSOCIATED
WITH *OLD*

u	40	50	60	70	80	90	100
Prob$_{(x)}$ (u)	0	0	0.2	0.2	0.2	0.2	0.2

These tables demonstrate some differences between the concepts and application of possibility and probability. It appears that the imprecision that is intrinsic in the natural language is possibilistic in nature and the possibility theory provides the basis for dealing with it in a systematic way.

VII. POSSIBILITY EXPECTED VALUE (PEV)

A. Possibility Measure

Further insight into the difference between possibility and probability can be gained by comparing the concept of a *possibility measure* with the frequently used concept of a probability measure. Let us first develop the concept of the possibility measure.

Definition 17: Let A be a nonfuzzy subset of U and X be a possibility distribution associated with a variable X which takes values in U. Then the *possibility measure* of A is denoted by p(A) and is defined as a number in [0.1] given by

$$p(A) = \operatorname*{Sup}_{u \in A} p_x(u), \tag{38}$$

where $p_X(u)$ is the possibility distribution function of Π_X. The interpretation of p(Z) may be as the possibility that a value of X belongs to Z, that is

$$\text{Poss } \{X \in A\} \triangleq p(A)$$
$$= \operatorname*{Sup}_{u \in A} p_X(u) \tag{39}$$

If A is a fuzzy set, then the concept of a value of X belonging to A is not meaningful. The following definition extends Equation 39 to the case of fuzzy sets.

Definition 18: Let A be a fuzzy subset of U characterized by a membership function m_A and let Π_X be a possibility distribution associated with a variable X which takes values in U. The *possibility measure*, p(A), of A is defined by

$$\text{Poss } \{X \text{ is } A\} \triangleq p(A)$$
$$= \operatorname*{sup}_{u \in U} m_A(u) \, \& \, p_X(u) \tag{40}$$

Note that "X is A" replaces "X ∈ A" in. It can be observed that the possibility measure may be defined in terms of the height of a fuzzy set as follows:

$$p(A) = hgt(A \cap \Pi_x) \qquad (41)$$

Example 7: As a simple illustration, consider the possibility distribution induced by the proposition "X is a small integer",

$$\Pi_x = 1/10 + 1/1 + .9/2 + .7/3 + .5/4 + .2/5$$

(1) Let A = {3,4,5,}. Then

$$p(A) = .7 \# .5 \# .2 \# = .7$$

(2) Let A be a fuzzy set of integers which are NOT SMALL,

$$A = .1/2 + .3/3 + .5/4 + .8/5 + 1/6 + ...$$

Then,

$$\text{Poss } \{X \text{ is not a small integer}\} =$$
$$hgt(.1/2 + .3/3 + .5/4 + .2/5) = .5 \qquad (42)$$

Notice that Equation 42 follows from the assertion that X is "a small integer" → Poss {X is "not a small integer"} or in symbols,

$$X \text{ is } F \rightarrow \text{Poss } \{X \text{ is } A\} = hgt(F \cap A)$$

which is implied by Equation 31 and 41. As a consequence, if A is a normal fuzzy set (i.e., hgt(A) = 1), then:

$$X \text{ is } A \rightarrow \text{Poss } \{X \text{ is } A\} = 1.$$

Let us now consider *composition of possibility measures*. The following relations based on the corresponding definitions of fuzzy set operations.

Let A and B be arbitrary fuzzy subsets of U. Then it follows from Equation 40 that the *possibility measure of the union of A and B* is given by

$$p(A \cup B) = p(A) \# p(B) \qquad (43)$$

By comparison, the corresponding relation for probability measures of A, B and A∪B (if they exist) is

$$p(A \cup B) \leq P(A) + P(B) \qquad (44)$$

and, if A and B are disjoint (i.e., $M_A(u)$ & $m_B(u) = 0$), then

$$P(A \cup B) = P(A) + P(B), \qquad (45)$$

which is the known additive property of probability measures. However, the possibility measure is not additive. Instead, it has the property expressed by Equation 43, which is analogous to Equation 45 with + replaced by #.

In a similar fashion, the *possibility measure of the intersection of A and B* is related to those of A and B by

$$p(A \cap B) \leqslant p(A) \text{ \& } p(B). \tag{46}$$

In particular, if A and B are *noninteractive*, i.e.,

$$m_{A \cap B} = m_A \text{ \& } m_B, \tag{47}$$

then (1.46) holds with the equality sign:

$$p(A \cap B) = p(A) \text{ \& } p(B). \tag{48}$$

By comparison, in the case of probability measures, the relation is

$$P(A \cap B) \leqslant P(A) \text{ \& } P(B) \tag{49}$$

and

$$P(A \cap B) = P(A) \cdot P(B) \tag{50}$$

if A and B are independent and nonfuzzy. Similarly as in case of Equations 43, Equation 48 is analogous to Equation 50 with product corresponding to & (min).

Possibility comes into play when given a feature i we ask, "How well is the constraint *feature i belongs to class M* satisfied when i is assigned to Y?" This question will be asked for each of the N classes, which will give us a possibility distribution for each feature i. Then same manipulations will be made to find the best M ($1 \geqslant M \geqslant N$) for the proposition *feature i belongs to class M*. After this is done for all features, a classification can be made based on criteria to be defined.

With possibility distributions comes the natural question of how to describe a distribution or set of distributions as an entity. The answer to that question is to define a Possibilistic Expected Value (PEV), for the distribution(s).

When working with the basket weaver and the kind of possibility distributions that Zadeh describes, we see reasons to ignore 0 possibilities for a basket weaver to make 100 baskets in a day when we get a PEV for the number of baskets he *can* make a day.

It also seems that we should give a higher weight to the greater possibilities, since anything associated with them is much more likely to occur than anything associated with a small possibility. For example if the distribution

$$\frac{.9}{2} + \frac{.1}{2}$$

occurred; the possibility of the distribution should be closer to 0.9, since the value associated with it is most likely to occur. It should be closer to 0.9 than the average of the two possibilities would be.

Definition 19: Given a possibility distribution

$$X_1/W_1 + X_2/W_2 + \ldots + X_3/W_3$$

the possibilistic expected value of the distribution is

$$PEV = \frac{\sum\limits_{i=1}^{N} X_i^2}{\sum\limits_{i=1}^{N} X_i} \qquad (51)$$

This definition will give us a number between 0 and 1 with the desired properties. We would like to be able to say we have a PEV with an associated W for a distribution. We therefore need a way to find W.

Definition 20: Given a possibility distribution and a possibilistic expected value (from 20) the W associated with that PEV is

$$W = \frac{\sum\limits_{i=1}^{N} X_i W_i}{\sum\limits_{i=1}^{N} X_i} \qquad (52)$$

The W is heavily influenced by the size of each possibility. It will tend to be close to the W_is that are associated with the larger possibilities. The combination of the PEV and W, written as PEV/W, of a distribution will give an indication of what value to expect from the distribution and how possible that value is (how good an indication it is).

For an example of the use of these definitions one can look at the case of a grocer who wants to know how many eggs to buy each day. He may have the following distribution for his town:

$$1/100 + 1/200 + .7/200 + .5/400 + .2/600 + .1/800$$

Then the PEV for the distribution will be

$$\frac{(1)^2 + (1)^2 + (.7)^2 + (.2)^2 + (.1)^2}{1 + 1 + .7 + .5 + .2 + .1} = \frac{2.79}{3.5} = .797$$

W will be

$$\frac{1 \times 100 + 1 \times 200 + .7 \times 300 + .5 \times 400 + .2 \times 600 + .1 \times 800}{1 + 1 + .7 + .5 + .2 + .1}$$

$$W = 910/3.5 = 260$$

We get, as a representation of our distribution, 0.797/260 from definitions 19 and 20. This can be interpreted as meaning that there is a 0.797 possibility that 260 eggs will be eaten in one day in the grocer's town. The grocer may then decide to use this information to buy 300 eggs per day.

Some other examples of distributions and their PEV and W's are

$$.3/1 + .1/2 + .3/4 + .4/5 + .4/6 + .8/7 + .3/8 + 0/9, .46/5.33$$

$$1/10 + .5/100 + .3/200 + .1/300, .71/79.95.$$

The possibility in the numerator gives an indication of how strong a measure the W is. The techniques of definitions 19 and 20 can also be used to describe several related possibility distributions. It will indicate what should be expected from them as a whole. The definitions must be extended a little to describe several distributions with one value. This fact has no real bearing on the basic thrust of this thesis and is merely noted.

Consider the fact that a 0 possibility for an W_i will not change the PEV of the distribution without the 0 possibility. This is when one is talking about basket weaving and you don't care whether Jack can weave 100 baskets in a day. In this case the higher possibilities are the important ones.

When you have a feature and are comparing a feature against each of N classes a 0 possibility does give some information. It indicates that the conceivability of the feature belonging to that class is negligible. That fact should be taken into account when the PEV is computed.

Now consider the following two distributions which result from the attempt to classify a feature vector of three features as belonging to one of two classes.

$$.8/\text{class } 1 + .7/\text{class } 1 + 0/\text{class } 1, \text{ PEV} = .7533$$

and

$$.83/\text{class2} + .72/\text{class2} + .5/\text{class2}, \text{ PEV} = .7286$$

Clearly, if the described object belongs to one of these two classes it belongs to the second one, since the possibility of each feature is higher for class 2. The third feature certainly does not belong to class 1. However, if we try to classify it by taking the distribution with the higher possibilistic expected value to be the one we want, which is a logical choice, the object will be classified as belonging to class 1, when in fact it should be classified as class 2.

VIII. N-ARY POSSIBILITY DISTRIBUTIONS

So far it has been assumed that in the translation of a proposition of the form

$$p \triangleq X \text{ is } F \rightarrow R(A(X)) = F \tag{53}$$

or equivalently,

$$X \text{ is } F \rightarrow \Pi_X = F, \tag{54}$$

p contains a single attribute $Z(X)$ whose possibility distribution is given by $\Pi_{A(X)} = F$. It is desirable to generalize the concept of the possibility distribution to such cases where p may contain n implied attributes $A_1(X), \ldots, A_n(X)$, with $A_i(X)$ taking values in U_i, $i = 1, \ldots, n$ (not necessarily distinct). It follows, that the translation of $p = X$ is F, where F is a fuzzy relation in the Cartesian product $U - U_1 \times \ldots \times U_n$, assumes the form

$$X \text{ is } F \rightarrow R(A_1(X), \ldots, A_n(X)) = F \tag{55}$$

or, equivalently,

$$X \text{ is } F \rightarrow \Pi(A(X), \ldots, A(X)) = F \tag{56}$$

where $R(A_1(X), \ldots, A_n(X))$ is an n-ary fuzzy restriction and $\Pi_A(X), \ldots, A(X))$ is an *n-ary possibility distribution induced by p*. Hence, the *n-ary possibility distribution function induced by p* is given by

$$p_A(X),...,A(X))(u_1,...,u_n) = m_F(u_1,...,u_n), \ (u_1,...,u_n) \in U \tag{57}$$

where m_F is the membership function of F.

Now consider a case where F is a Cartesian product of n unary fuzzy relations, $F_1, \ldots, F_n$, i.e., $F = F_1 \ x \ldots x \ F_n$. Hence, the right-hand side of Equation 53 decomposes into a system of n unary relational assignment equations, i.e.,

$$X \text{ is } F \rightarrow R(A_1(X)) = F_1$$

$$R(A_2(X)) = F_2$$

$$\vdots$$

$$\vdots$$

$$R(A_n(X)) = F_n$$

$$\tag{58}$$

Correspondingly,

$$(A_1(X),...,A_n(X)) = A_1(X) \ x...x \ A_n(X) \tag{59}$$

and

$$p_{A_1}(x),...,A_n(x)) \ (u_1,...,u_n) = p_{A_1}(x)(u_1),...,p_{A_n}(x)(u_n) \tag{60}$$

where

$$p_{A_i}(x)(u_i) = m_{F_i}(u_1), \ u_i \in U, \quad i = 1,...,n. \tag{61}$$

IX. APPLICATION TO A COMPUTER SECURITY MODEL

With the advent of the computer as a major tool in our society there has developed a new class of crime which takes advantage of the fact that the introduction of the computer into an organization creates changes in the security procedures and requirements of that organization which are not yet fully understood. The fact that a computer is both faster and more efficient than a person in carrying out its tasks makes the changes even more significant. Where the unnoticed embezzlement of a large sum of money in a totally unautomated accounting system might take years, the use of the computer by a skilled programmer could cut the time for such a crime down to weeks or even days. The fact that large computer systems are usually so complex as to defy total understanding also complicates the task of protection of assets. If you combine this with the realization that the computer criminal is often extremely knowledgeable about the system whose security he or she is compromising, the size of the problem becomes clear.

According to Pritchard, there are three types of loss which an organization does not want its computing system to suffer. These are loss of availability, loss of integrity and loss of confidentiality.[15] Thus the security measures which an organization implements for its computing system should be designed to make it impossible for any of these types of loss to occur. However, it is the nature of security that absolute security is unattainable. No matter how well designed and implemented a security system is, there will always be some means of compromising the system's availability, integrity, or confidentiality. As a result of this, the purpose of a security system given this restraint is to minimize the probability of exposure to any of these types of loss.[13]

One generalized model which has been used with some success in the preparation of a program for designing and modifying security programs for a computing system is the risk analysis model. Martin,[13] Pritchard,[15] and Farr et al.[3] have all written quite extensive volumes discussing the use of this model is designing security systems. However, all of their models are based upon the concept of hard probability which for previously mentioned reasons is not adequate in working with generalized system models. It is this type of model, specifically the model presented in a paper by Brocks,[1] which shall be fuzzified in this chapter.

The risk analysis model is based upon two important presuppositions. First, absolute protection of a computing system is unobtainable. Second, the amount of money which any organization is able to spend in pursuit of protection is limited.[13] It is necessary as a result of these presuppositions to develop a method for determining which set of security measures will maintain the highest level of cost effectiveness in the security system.

Definition 21: A threat is defined to be any action or event whose occurrence would adversely effect the computing system.[2] Another term for threat is hazard.

Definition 22: The vulnerability of a system to a certain threat is defined to be the cost which the organization would incur if that threat takes place.[2] Other terms for vulnerability are loss expectancy and cost.

Definition 23: The risk of a system for a certain threat is defined to be the vulnerability of the system to that threat multiplied by the probability of the occurrence of that threat within a given period of time.[2] Another term for risk is exposure.

The risk analysis model as presented by Brocks list four separate stages in the development of a security program for a computing system. First, one must identify the threats to which the system is exposed. Second, one must define the vulnerability of the system to each of these threats and determine the probability of these events occurring. Third, one must select the most suitable protection mechanisms for each threat. Finally, one must implement and monitor the protection mechanisms.[1]

The first step of this model, identification of the threats to the system, involves attempting to list every possible situation which would cause the compromise of the system via loss of availability, loss of integrity or loss of confidentiality. This step has been studied quite thoroughly. Even though every organization will have its own unique set of threats, there are many threats which are common to all systems. Lists of these common threats have been compiled in a number of volumes.[3-8,11,13-17]

The second step in Brocks' model, the assessment of the amount of risk created by each threat, is where the actual risk analysis of the system begins. Brocks divides this step in three parts. First, management is asked to determine an amount which is to be considered the lowest "crippling loss" for the organization. Then, an index value between 1 and 100 is given to each threat. The specific index value computed for any threat is defined to be 100 times the vulnerability of the system to the threat divided by the organization's crippling loss.[1] An example of this would be a situation where a corporation's vulnerability to a certain threat is $300,000 and the corporation's crippling loss is $1 million. This threat would receive a vulnerability index of 30. Finally, an estimate is made as to the probability of occurrence of each threat. This estimate is to be a subjective evaluation of the likelihood of the occurrence of the threat made on the basis of the frequency of the system's exposure, the relative hostility of the environment, and past experience.[1]

The third step in Brocks' model, the choice of suitable protection mechanisms, is similar to the second. The management of the organization in this step sets a minimum "crippling cost" to the organization for providing protection against each threat. Each of the possible protection mechanisms for a threat is given a cost index by the formula 100 times the cost of the mechanisms divided by the crippling cost.[1]

The protection mechanisms are evaluated by the comparison of the mechanism cost index with the "expected value of loss" index which is obtained by multiplying the vulnerability index by the probability index. Through the comparison of the indices, management can decide which protection mechanisms are cost effective in terms of the organizations security requirements and which are not.[1]

The fourth step of Brocks' model, implementation and monitoring of the protection mechanisms, simply stresses the obvious fact that the situation in which security for a computing system is being provided is always changing. As a result of this, there must be an organized and ongoing effort to maintain the optimal level of security possible based upon the changing cost restraints and security requirements of the system.

There are several changes which could be made to improve Brocks' risk analysis model. First, Brocks admits that many of the probabilities required by his model are soft probability estimates. In these situations, the fuzzy expected value could be used in a group decision making environment to give a better estimate of the soft expected value. A second weakness is that there is no method given in Brocks' model for handling the varying amounts of protection which different mechanisms provide for the same threat. A final weakness is that there is no definitive method for management to use in making a decision as to what set of security measures will provide the maximum amount of protection for the system as a whole.

If one creates a fuzzy set MEASURES WHICH PROVIDE FULL PROTECTION AGAINST THE THREAT for each threat, then an application of the typical value of a population of possibility distributions algorithm which was developed in the previous chapter provides a method for choosing the best set of protection mechanisms in a group decision making environment. These techniques will now be applied to the risk analysis model to give a fuzzified version of the model.

The first step in the fuzzy risk analysis model is the identification of threats to the system. It is obvious that any and all models of security must have some knowledge of what threats they are attempting to deter. Thus, the first step of the new model will consist of the determination of a set of n threats T with a generic element of T being denoted by the symbol T_i, $i \leq n$.

The second step in the model is the assessment of the risk which each threat induces. This step can be broken into four distinct parts.

The first part of the risk assessment step is the determination of the vulnerability of the system to the specific threat being considered. This can be done in two ways. First if empirical evidence of the cost of recovering from the threat is available, it should be used. An example of this would be the estimation of the vulnerability of the computing center to fire by the use of the present replacement cost of all of the equipment in the computing center. The second technique can be used if no empirical evidence of vulnerability is available. This technique is to take soft estimates of the vulnerability to the threat from the members of the group creating the security program. The estimates will fall between some upper and lower bound. A typical estimate of vulnerability can be computed by finding the fuzzy expected value of the estimates in the interval between the lower bound and the upper bound. The set of vulnerability estimates is called V with a generic element of V being denoted by the symbol V_i.

The second part of the risk assessment step is the estimation of a probability of occurrence for each threat T_i. The set of probability estimates is called P with a generic element of P being denoted by P_i. Brocks states that the probability estimates which are used in his model are for the most part subjective evaluations. In the fuzzy model, there are two ways for estimating a member of P. First, if empirical evidence is available, use it to get a hard probability estimate. Second, if a subjective estimate of probability is required, have each member of the computer security group assign a membership value for the threat in the fuzzy set EVENTS WHICH WILL OCCUR DURING THE SPECIFIED TIME PERIOD. The typical value of probability for that threat will be the fuzzy expected value of each of the membership values for the threat as given by the group.

The third part of the risk assessment step is the computation of the organization's risk for each specific threat. The set of risks is called R with a generic element of R being denoted by R_i. The risk R_i of the organization to a threat T_i is computed by the formula $R_i = V_i \times P_i$. R_i represents the amount which the organization could reasonably expect to lose if no protection is provided against threat T_i.

The final part of the risk assessment step is the computation of a set of priority percentages PP corresponding to the set of threats T. The priority percentages are found according to the formula $PP_i = R_i / \sum_{j=1}^{n} R_j$. The priority percentages represent the relative amount of damage which can be expected from each threat when compared to the damage which can be expected from all of the threats.

The third step in the fuzzy risk analysis model is the determination of a set of most effective protection mechanisms for the set of threats confronting the system. This step can be divided into five distinct parts.

The first part of the mechanism selection step is the setting of the maximum amount which the organization can spend to provide security against all of the threats. This amount is called TOTAL.

The second part of the mechanism selection step is the creation of a list of possible protection mechanisms for each threat. This list should be created by consulting with staff members who have knowledge of the system and by studying the literature for suggested mechanisms. The list of mechanisms created for any threat T is called M_i. An element of M_i shall be denoted by M_{ij}.

The set M_i should include as single elements each mechanism by itself, mechanisms which can be used in combination with each other and the null mechanism. A possible set of mechanisms for the protection of a terminal room from unauthorized entry might be (1) posting a guard at the door; (2) providing new locks for the door; (3) using a closed circuit television system to record access to the terminal room; (4) posting a guard and providing new locks for the door; (5) posting a guard and using closed circuit television; (6) providing new locks and using closed circuit television; (7) posting a guard, using closed circuit television and providing new locks for the door; and (8) the null mechanism.

The third part of the mechanism selection step is to have each person in the security group estimate a compatibility value, $\chi (M_{ij})$, which represents the compatibility of mechanism M_{ij} to the security needs of the organization against threat T_i. These compatibility values are to be assigned in the interval [0,1] with 1 representing the fact that mechanism M_{ij} will fully protect the organization against threat T_i, 0 representing the fact that mechanism M_{ij} provides no protection against threat T_i, and values between 0 and 1 representing partial protection against the threat. Each set of mechanisms M_i along with its corresponding set of compatibility values are a fuzzy set MECHANISMS WHICH PROVIDE FULL PROTECTION AGAINST THE THREAT T_i. For each mechanism M_{ij}, each of member of the security group should also estimate a cost C_{ij} for that mechanism. This estimate can either be hard or soft depending upon the availability of empirical evidence as to the cost. If we allow a variable X_i to take as its universe the mechanisms in set M_i, the assignment X_i is restricted by the fuzzy set MECHANISMS WHICH PROVIDE FULL PROTECTION AGAINST THE THREAT T_i creates a possibility distribution for X_i. This possibility distribution gives the possibility that each mechanism in M_i will fulfill the security requirements for the threat T_i.

The fourth part of the mechanism selection step is the conjunction of the possibility distributions for each X_i to give a possibility distribution for the n-tuples of mechanisms created by taking the Cartesian product of the sets M_i. This can be classified as the conjunction of several related restrictions upon n variables. The restrictions are related for three reasons. First, the amount of security which each mechanism provides to the system as a whole is different. Second, the amount of security which the system as a whole receives is the sum of the security which each of the mechanisms provide. Finally, there is a threshold cost which the n-tuple of mechanisms to be chose can not exceed.

For each member of the computer security group, there should be n possibility distributions corresponding to the n threats to the system. Each threat T_i should have a priority percentage PP_i which was computed earlier. Each mechanism M_{ij} in the set of possibility distributions should have a cost C_{ij} and a possibility value $Poss(M_{ij})$ which was created by the assignment of the fuzzy set MECHANISMS WHICH PROVIDE FULL PROTECTION AGAINST THE THREAT T_i as a restriction upon the variable X_i. The fuzzy relation which we will use in this model to give a possibility value for each ordered n-tuple of mechanisms which can be combined to provide security against the n threats is the formula

$$Poss(M_{ij},\ldots,M_{nj}) = \begin{cases} 0 & \text{if } \sum_{i=1}^{n} C_{ij} > \text{Total} \\ \sum_{i=1}^{n} PP_i \times Poss(M_{ij}) & \text{if otherwise} \end{cases}$$

The application of this fuzzy relation to every n-tuple in the Cartesian product of the sets M_i creates a possibility distribution representing the possibility that those n-tuples will meet the protection requirements of the system against the set of threats, while not exceeding the total allowable expenditure for security. At this point in the model there should be one possibility distribution for each person in the computer security group.

The final part of the mechanism selection step is the application of the typical value of a population of possibility distributions algorithm created in the previous chapter of the possibility distributions of the computer security group. The result of this algorithm will be the set of mechanisms which the group as a whole has chosen as both providing a maximum amount of protection while also staying within the security budget of the organization.

The final step in the fuzzy risk analysis model is the implementation and monitoring of the security mechanisms chosen. This step is the same as in Brocks' model. The only addition which this model makes is the requirement that the security group meet at set intervals of time to reevaluate the protection mechanisms selected.

As the final subject of this chapter, let us consider the following application of the fuzzy risk analysis model to the design of a security program in a hypothetical situation. A synopsis of the entire model is given in Table 4 to make reference to the steps in the model easier. It should be noted that the mechanism cost estimates used in this example were chosen arbitrarily and may not reflect the actual costs of the corresponding mechanisms in the real world.

An organization has decided to add a new terminal room in its plant for the use of staff members in two confidential project groups. This terminal room contains ten terminals which are to be used by staff members working on the first project and six terminals which are to be used by the staff of the second project. The organization desires to provide security for the room, but must do so within the maximum cost of $600 per year.

Table 4
STEPS IN THE FUZZY RISK
ANALYSIS MODEL

1. Identify set of threats T
2. Assess risk of system to each threat T_i
   a. Estimate vulnerability V_i
   b. Estimate probability of occurrence P_i
   c. Compute risk R_i
   d. Compute priority percentage PP_i
3. Select appropriate set of security mechanisms
   a. Determine maximum amount to be spent on security
   b. Identify possible protection mechanisms M_{ij}
   c. Estimate compatibility (M_{ij}) and cost C_{ij}
   d. Take conjunction of possibility distributions
   e. Apply typical value of a population of possibility distributions algorithm to choose n-tuple of mechanisms
4. Implement and monitor mechanisms chosen

The first step in the fuzzy model is the determination of threats to the terminal room. The security group of three members lists three elements in set T. These are (1) a terminal may be stolen or damaged intentionally, (2) a fire may destroy the terminal room, and (3) a staff member of one of the projects may attempt to use a terminal which is to be used only by staff members of the other project.

The second step of the fuzzy model is the assessment of risk. The members of the security group estimate V_1 and V_2 on the basis of empirical data, which in this case is the replacement cost of the terminals. Thus $V_1 = \$1000$ and $V_2 = \$16,000$. No empirical data is available on the vulnerability of the organization to threat T_3. To obtain a vulnerability value, each member of the group estimates the vulnerability. The estimates are $600, $520, and $650. The group estimate for V_3 is the fuzzy expected value of the estimates in the interval [520, 650]. Thus $V_3 = \$600$.

The second part of the risk assessment step is the estimation of the probability of occurrence for each threat T_i. In this case, no empirical evidence is available for any of the threats. As a result of this, the members of the security group each estimate what he or she believes the probability of each threat's occurrence might be. The estimates for P_1 are 0.05, 0.06, and 0.02. The estimates for P_2 are 0.01, 0.02, and 0.04. The estimates for P_3 are 0.90, 0.70, and 0.50. To find a typical estimate of the probability of the occurrence of each threat, the fuzzy expected value of each set of estimates in the interval [0,1] is computed. The estimates of the security group are 0.06 for P_1, 0.04 for P_2 and 0.70 for P_3.

The third part of the risk assessment step in the model is the computation of the risk for each threat. Using the formula $R_i = V_i \times P_i$, the risks are computed to be $R_1 = \$60$, $R_2 = \$720$, and $R_3 = \$420$.

The final part of the risk assessment step is the computation of the priority percentages. For this example PP_1 is computed to be 0.05, PP_2 is computed to be 0.60 and PP_3 is computed to be 0.35.

The first part of the mechanism selection step is the creation of lists of possible protection mechanisms for each threat. After studying the literature, the security group decides upon the protection mechanisms are given in Table 5.

For the rest of this example, the mechanisms described in Table 3 shall be specified by the combination of the mechanism set number i (where this is not clear from the context) and the mechanism number j.

Table 5
THREATS T AND PROPOSED COUNTER MEASURES M_i

T_1 Stolen terminal
$\quad$ M_{11} Bolting the terminals to the tables
$\quad$ M_{12} Installing a new lock on terminal room door
$\quad$ M_{13} Bolting the terminals to the tables and installing a new
$\qquad$ lock on terminal room door
$\quad$ M_{14} The null mechanism
T_2 Fire in terminal room
$\quad$ M_{21} Installing a sprinkler system
$\quad$ M_{22} Installing fire extinguishers
$\quad$ M_{23} The null mechanism
T_3 Unauthorized use of terminal
$\quad$ M_{31} Using badge system to activate terminals
$\quad$ M_{32} Using passwords to limit access to terminals
$\quad$ M_{33} The null mechanism

Table 6
ESTIMATED COMPATIBILITY OF MECHANISM TO THREAT

First member of security group
$\quad (M_1) = 0.75/1 + .5/2 + 0.8/3 + 0.4$
$\quad (M_2) = 0.8/1 + 0.4/2 + 0/3$
$\quad (M_3) = 0.4/1 + 0.3/2 + 0/3$
Second member of security group
$\quad (M_1) = 0.6/1 + 0.6/2 + 0.8/3 + 0/4$
$\quad (M_2) = 0.8/1 + 0.2/2 + 0/3$
$\quad (M_3) = 0.5/1 + 0.4/2 + 0/3$
Third member of security group
$\quad (M_1) = 0.4/1 + 0.4/2 + 0.7/3 + 0/4$
$\quad (M_2) = 0.7/1 + 0.5/2 + 0/3$
$\quad (M_3) = 0.6/1 + 0.7/2 + 0/3$

The second part of the mechanism selection step is the assignment of grades of membership in the set MECHANISMS WHICH PROVIDE FULL PROTECTION AGAINST THE THREAT for each mechanism in M_i. The compatiblity values as estimated by each member of the security group are given in Table 6.

Along with defining the fuzzy sets, each member of the security group also estimates the cost of each of the protection mechanisms. Three of these estimates are based upon hard data which were obtained from manufacturers. These include the sprinkler system, whose cost is $6000, the fire extinguishers, whose cost is $200, and the badge system, whose cost is $150. (Remember that these cost figures are arbitrarily chosen for this example and may not reflect the actual cost of such mechanisms.) The cost estimates of the group are given in Table 7.

After the estimation of the grades of membership in the sets and the estimation of the costs of the mechanisms, the remaining steps in selecting the proper security mechanisms are algorithmic in nature. The security group uses the fuzzy relation defined earlier in this chapter to take the conjunction of the three related possibility distributions corresponding to the three threats. This gives a single possibility distribution corresponding to the protection provided to the system by the n-tuples of mechanisms for each person in the security group are given in Tables 8 to 10.

The final part of the mechanism selection step is the application of the typical value of a population of possibility distributions algorithm to the three distributions given in the preceding paragraphs. In this case, the typical value of the distribution for security group

Table 7
ESTIMATED COST OF
EACH MECHANISM

First member of security group
C_1 = $25/1 + $50/2 + $75/3 + $0/4
C_2 = $6000/1 + $200/2 + $0/3
C_3 = $150/1 + $300/2 + $0/3
Second member of security group
C_1 = $30/1 + $40/2 + $75/3 + $0/4
C_2 = $6000/1 + $200/2 + $0/3
C_3 = $150/1 + $400/2 + $0/3
Third member of security group
C_1 = $25/1 + $55/2 + $70/3 + $0/4
C_2 = $6000/1 + $200/2 + $0/3
C_3 = $150/1 + $160/2 + $0/3

Table 8
POSSIBILITY DISTRIBUTION FOR
FIRST MEMBER OF THE GROUP

N-Tuple	Possibility	N-Tuple	Possibility
(1,1,1)	0.0000	(3,1,1)	0.0000
(1,1,2)	0.0000	(3,1,1)	0.0000
(1,1,3)	0.0000	(3,1,1)	0.0000
(1,2,1)	0.4175	(3,2,1)	0.4200
(1,2,2)	0.3825	(3,2,2)	0.3850
(1,2,3)	0.2775	(3,2,3)	0.2800
(1,3,1)	0.1775	(3,3,1)	0.1800
(1,3,2)	0.1425	(3,3,2)	0.1450
(1,3,3)	0.0375	(3,3,3)	0.0400
(2,1,1)	0.0000	(4,1,1)	0.0000
(2,1,2)	0.0000	(4,1,2)	0.0000
(2,1,3)	0.0000	(4,1,3)	0.0000
(2,2,1)	0.4050	(4,2,1)	0.3800
(2,2,2)	0.3700	(4,2,2)	0.3450
(2,2,3)	0.2650	(4,2,3)	0.2400
(2,3,1)	0.1650	(4,3,1)	0.1400
(2,3,2)	0.1300	(4,3,2)	0.1050
(2,3,3)	0.0250	(4,3,3)	0.0000

members one and two is the 3-tuple (3,2,1). The typical value for group member three is the 3-tuple (3,2,2). Thus the counter for 3-tuple (3,2,1) is set at 2 and the counter for 3-tuple (3,2,2) is set to 1. Counters for all the other 3-tuples are set to 0. Since the 3-tuples of mechanisms are not intrinsically numeric events, it is necessary that each 3-tuple be arbitrarily assigned a compatiblity value which in this case will be an integer between 1 and 36. The arbitrary assignment in this case will be in reverse lexigraphic order. After the scaled values for $\mu\star(\zeta_T\star)$ are computed from the counters, the fuzzy expected value of the set of 3-tuples is computed in the interval [1,36]. The compatibility value and $\mu\star(\zeta_T\star)$ for each 3-tuple is given in Table 11.

The fuzzy expected value of this set of 3-tuples is 15. Thus, the set of mechanisms which in the opinion of the security group gives the best protection to the terminal room while at the same time meeting the security budget is the set represented by the 3-tuple (3,2,1).

The final step in the fuzzy risk analysis model is the implementation and monitoring of the measures. To provide protection for the terminal room, the organization takes the fol-

Table 9
POSSIBILITY DISTRIBUTION FOR
SECOND MEMBER OF THE GROUP

N-Tuple	Possibility	N-Tuple	Possibility
(1,1,1)	0.0000	(3,1,1)	0.0000
(1,1,2)	0.0000	(3,1,2)	0.0000
(1,1,3)	0.0000	(3,1,3)	0.0000
(1,2,1)	0.3250	(3,2,1)	0.3350
(1,2,2)	0.0000	(3,2,2)	0.0000
(1,2,3)	0.1500	(3,2,3)	0.1600
(1,3,1)	0.2050	(3,3,1)	0.2150
(1,3,2)	0.1700	(3,3,2)	0.1800
(1,3,3)	0.0300	(3,3,3)	0.0400
(2,1,1)	0.0000	(4,1,1)	0.0000
(2,1,2)	0.0000	(4,1,2)	0.0000
(2,1,3)	0.0000	(4,1,3)	0.0000
(2,2,1)	0.3250	(4,2,1)	0.2950
(2,2,2)	0.0000	(4,2,2)	0.2600
(2,2,3)	0.1500	(4,2,3)	0.1200
(2,3,1)	0.2050	(4,3,1)	0.1750
(2,3,2)	0.1500	(4,3,2)	0.1400
(2,3,3)	0.0300	(4,3,3)	0.0000

Table 10
POSSIBILITY DISTRIBUTION FOR
THIRD MEMBER OF THE GROUP

N-Tuple	Possibility	N-Tuple	Possibility
(1,1,1)	0.0000	(3,1,1)	0.0000
(1,1,2)	0.0000	(3,1,2)	0.0000
(1,1,3)	0.0000	(3,1,3)	0.0000
(1,2,1)	0.5300	(3,2,1)	0.5450
(1,2,2)	0.5650	(3,2,2)	0.5800
(1,2,3)	0.3200	(3,2,3)	0.3350
(1,3,1)	0.2300	(3,3,1)	0.2450
(1,3,2)	0.2650	(3,3,2)	0.2800
(1,3,3)	0.0200	(3,3,3)	0.0350
(2,1,1)	0.0000	(4,1,1)	0.0000
(2,1,2)	0.0000	(4,1,2)	0.0000
(2,1,3)	0.0000	(4,1,3)	0.0000
(2,2,1)	0.5300	(4,2,1)	0.5100
(2,2,2)	0.5650	(4,2,2)	0.5450
(2,2,3)	0.3200	(4,2,3)	0.3000
(2,3,1)	0.2300	(4,3,1)	0.2100
(2,3,2)	0.2650	(4,3,2)	0.2450
(2,3,3)	0.0200	(4,3,3)	0.0000

lowing steps: (1) the terminals are bolted to the tables, (2) new locks are placed on the terminal room doors, (3) fire extinguishers are placed in the terminal room, and (4) a badge system is installed to limit access to the use of the terminals. As a final step, the security group agrees to review the success of the protection mechanisms chosen every 3 months.

Although this example includes only a small number of threats and protection mechanisms, the fuzzy model can be used with the same degree of effectiveness in situations which have more complex sets of threats and mechanisms.

Table 11
FUZZY EXPECTED VALUE INFORMATION

3-Tuple	χ(X,Y,Z)	$\mu\star$ ($\zeta_T\star$)	3-Tuple	χ(X,Y,Z)	$\mu\star$ ($\zeta_T\star$)
(1,1,1)	36	1.00	(3,1,1)	18	1.00
(1,1,2)	35	1.00	(3,1,2)	17	1.00
(1,1,3)	34	1.00	(3,1,3)	16	1.00
(1,2,1)	33	1.00	(3,2,1)	15	1.00
(1,2,2)	32	1.00	(3,2,2)	14	24.33
(1,2,3)	31	1.00	(3,2,3)	13	36.00
(1,3,1)	30	1.00	(3,3,1)	12	36.00
(1,3,2)	29	1.00	(3,3,2)	11	36.00
(1,3,3)	28	1.00	(3,3,3)	10	36.00
(2,1,1)	27	1.00	(4,1,1)	9	36.00
(2,1,2)	26	1.00	(4,1,2)	8	36.00
((2,1,3)	25	1.00	(4,1,3)	7	36.00
(2,2,1)	24	1.00	(4,2,1)	6	36.00
(2,2,2)	23	1.00	(4,2,2)	5	36.00
(2,2,3)	22	1.00	(4,2,3)	4	36.00
(2,3,1)	21	1.00	(4,3,1)	3	36.00
(2,3,2)	20	1.00	(4,3,2)	2	36.00
(2,3,3)	19	1.00	(4,3,3)	1	36.00

X. CONCLUSIONS

In this study, we have presented a risk analysis model of computer security which uses fuzzy set theory. This model has three distinct advantages over the probabilistic risk analysis model as presented by Brocks.[1] First, there exists in the fuzzy model a method for selecting a set of protection mechanisms for the system. Brocks' model, after providing those making the selection with the basic tools of risk analysis, does not give any solid algorithm for choosing the protection mechanisms. Second, the use of the fuzzy expected value in a group decision making environment to determine soft probability gives a much more central and thus reliable estimate of probability to use in the risk calculations. Finally, Brocks' model does not address the question of the relationship between the amount of protection provided by a mechanism and the amount of protection needed to make the system totally secure against a threat. The use of fuzzy sets to handle this problem makes it possible to add to the model what is without a doubt vital information in the system.

The validity of the fuzzy model presented has not been demonstrated by experimental or observational evidence. Comparing the effectiveness of the fuzzy model against the effectiveness of other models remains a problem for future study. Original experimentation in this area should probably consist of in-depth comparisons of the effectiveness of the set of protection mechanisms suggested by the fuzzy model and the set of protection mechanisms suggested by the probabilistic risk analysis model in identical situations. Due to the complexity of these models, these first experiments should probably be limited to systems whose security requirements are fairly obvious. For the application of the fuzzy model to complex systems, the development of a computer program which takes the conjunction of the possibility distributions would be extremely helpful. Since this process is algorithmic in nature, such a program would probably be fairly easy to write.

There are two areas where it may be possible to improve on the fuzzy risk analysis model. The first area is the fuzzy relation which is used in the model to take several possibility distributions for single mechanism and combine them into a single possibility distribution representing the effectiveness of n-tuples of mechanisms. It may be the case that a more effective function for this fuzzy relation could be found. The second area in which improvement may be possible is in the application of the typical value of a population of possibility distributions algorithm. Because the n-tuples of protection mechanisms are not numeric events, compatibility values for the n-tuples are arbitrarily assigned in the computation of the fuzzy expected value. It is possible that some method could be found to assign compatibility values which would be based upon some intrinsic quality of the n-tuples of protection mechanisms. These two areas remain open for further study.

The use of the fuzzy expected value and the typical value of a population of possibility distributions algorithm in the fuzzy risk analysis model is just one possible application of these methods to decision making. These methods can and should be applied to situations where decisions (particularly group decisions) must be made in an uncertain environment. Some possible areas of application of these methods are stock selection, urban planning, business decision making, and computer pattern recognition.

ACKNOWLEDGMENTS

This work was partially supported by NSF grant IST 8405953.

REFERENCES

1. **Brocks, B. J.**, *Security Problems in EDP — Assessing the Risk*, Proceedings of the 1974 Eurocomp Conference, 1974, 765.
2. **Fannin, D.**, *Guidelines for Establishing a Computer Security Program*, Massachussetts Computer Security Institute, Hudson, Mass., 1979.
3. **Farr, M. A. L., Chadwick, B., and Wong, K. K.**, *Security for Computer Systems*, National Computing Center, Manchester, England, 1972.
4. **Green, G. and Farber, R. C.**, *Introduction to Security*, Security World Publishing Co., Los Angeles, 1975.
5. **Hamilton, P.**, *Computer Security*, Associated Business Programs, London, 1972.
6. **Hemphill, C. F. and Hemphill, J. M.**, *Security Procedures for Computer Systems*, Dow Jones-Irwin, Homewood, Ill., 1973.
7. **Hoffman, L. J.**, *Security and Privacy in Computer Systems*, Melville Publishing, Los Angeles, 1973.
8. **Hsiao, D. K., Kerr, D. S., and Madnick, S. E.**, *Computer Security*, Academic Press, New York, 1979.
9. **Kandel, A.**, *Fuzzy Mathematical Techniques with Application*, Addison Wesley, Reading, Mass., 1986.
10. **Kandel, A.**, *Fuzzy Statistics and System Security*, Proc. 1980 Int. Conf: Security through Science and Engineering, West Berlin, 1980.
11. **Krauss, L. I.**, *SAFE: Security Audit and Field Evaluation for Computer Facilities and Information Systems*, AMACOM, New York, 1972.
12. **Margolis, I.**, *private communications*, 1981.
13. **Martin, J.**, *Security, Accuracy, and Privacy in Computer Systems*, Prentice-Hall, Englewood Cliffs, N.J., 1973.
14. **Parker, D. B.**, *Crime by Computer*, New York: Scribner, Inc., New York, 1976.
15. **Pritchard, J. A. T.**, *Risk Management in Action*, NCC Publications, Manchester, England, 1978.
16. **Walker, B. J. and Blake, I. F.**, *Computer Security and Protection Structures*, Dowden, Hutchingon & Ross, Stroudsburg, Pa., 1977.
17. **Wooldridge, S., Corder, C. R., and Johnson, C. R.**, *Security Standards for Data Processing*, John Wiley & Sons, New York, 1973.
18. **Zadeh, L. A.**, *Information and control, Fuzzy Sets Syst.* 8, 338, 1965.
19. **Zadeh, L. A.**, *Fuzzy sets as a basis for a theory of possibility, Fuzzy Sets and Systems*, 1, 3, 1978.
20. **Zadeh, L. A.**, Quantitative fuzzy semantics, *Information Sciences*, 3, 159, 1971.
21. **Zadeh, L. A.**, *Fuzzy Logic and Approximate Reasoning*, Memorandum No. ERL-M479, Electronic Research Lab., University of California, Berkeley, Calif, 1974.
22. **Zadeh, L. A.**, The concept of a linguistic variable and its application to approximate reasoning, Part I, *Inf. Sci.*, 8, 199, 1975; Part II, *Inf. Sci.*, 9, 43, 1971.

Chapter 3

STRUCTURE MODELING OF PROCESS SYSTEMS FOR RISK AND RELIABILITY ANALYSIS

Ahmad Shafaghi

TABLE OF CONTENTS

I. INTRODUCTION

Chemical process plants are complex integrated systems which often operate under limiting conditions. The raw materials used are flammable, toxic, corrosive, and explosive. Therefore, process plants are sources of hazards, and they potentially pose a risk to the public and plant employees.

Process system safety and reliability analysis is a systematic, rigorous decision-making approach to improve system reliability and subsequently to control the risk. This goal-oriented activity is normally carried out through three tasks, shown in Figure 1.

The first task is to identify potential hazards and operability problems. In general, there are two types of hazard: inherent, due to the nature of raw materials used; and subtle, due to omissions and errors made in design. Identifying these hazards is crucial, because the effectiveness of the other two tasks depends on it. The technique of Hazard and Operability (HAZOP) study[1] is a systematic, creative method for this task and has several advantages over more traditional methods, e.g., "what if".

After identification of the hazards, the second task is to estimate the risks associated with them. Risk is defined as both the likelihood of the hazardous event and the severity of the accident. The severity of the accident consequence is usually defined as the degree, sometimes in terms of frequency, of exposure to the accident. The system models, or fault trees, to be discussed in this paper are used during this second task to quantify the likelihood of hazardous events.

The third task, and the ultimate goal of modeling in process system safety and reliability analysis, is to control the risks. This final task is carried out by comparing the calculated risks with risk criteria specified by an authority. The criteria can be based on past experience or determined by the company's objectives. If the calculated risks are beyond the specified limit, changes can be made to improve the design or operation procedures to reduce the risk. This part of the task is shown by the feedback line in Figure 1.

The information yielded during the tasks shown in Figure 1 is valuable whether it is qualitative or quantitative. For quantitative results, reliablity data on the plant equipment and human interactions must be collected and analyzed for use with the system models. To obtain a realistic, quantitative reliability measure on the system, the data must be realistic.

II. MODELING FOR RISK ESTIMATION

During the early stages of a design, a preliminary hazard identification can identify inherent hazards due to the nature of materials used. A more rigorous hazard identification can be carried out at later stages of design to identify subtle hazards. The results of a hazard identification program is a set of potential hazardous events and operability problems.

However, a hazard identification program can only identify the hazards. The ways the hazards could be realized and propagated must still be specified. Fault trees, often complemented by event trees, are used for this purpose.

Development of fault trees for process systems which are in continuous operation is difficult, because there are so many ways for an operating system to fail. As a result, the construction of process system fault trees has become a major and independent activity, often called fault tree synthesis.

The relationship of fault tree construction to the overall task of risk calculation (Task 2) is depicted in Figure 2. After major hazards (or undesirable events) have been identified, fault trees are developed to focus on these events (often called top events). Fault trees are logic diagrams in which basic events (e.g., hardware failures, human errors, and environment impacts) are combined to lead to a particular top event. When analyzed using reliability data (e.g., equipment failure rates and human error probabilities), fault trees provide qualitative and more importantly, quantitative probabilistic information on the modes of system failure.

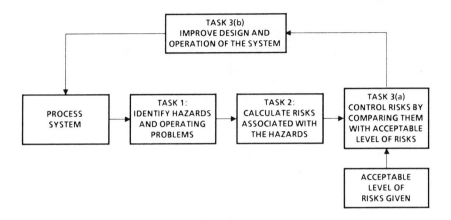

FIGURE 1. An overview of process system safety and reliability analysis.

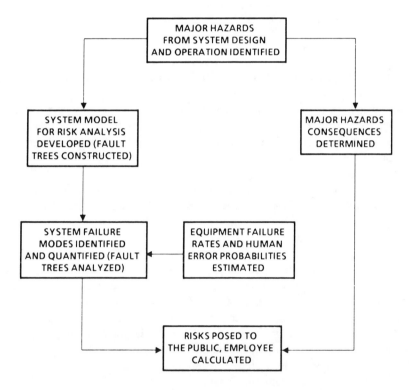

FIGURE 2. Systematic steps in risk assessment of process system.

Risk is calculated by determining both the likelihood and severity of system failure. There may be several types of system failure, each of which may result in a spectrum of consequences. Hence, risk is defined in terms of a system of triplets, i.e., the accident type, the probability of the accident, and the severity of the accident consequence. Decisions should be made based on the study of this system.

The method used for fault tree construction can significantly affect the remainder of the task because it represents the basic "view" of the plant. In component modeling, the plant is seen as a collection of components, while in structure modeling, the plant is seen as a system of control loop structures. These two methods of fault tree construction — component modeling and structure modeling — are discussed below.

Table 1
MAJOR METHODS OF FAULT
TREE SYNTHESIS BY COMPONENT
MODELING[2]

Component model form	Fault tree synthesis computer code names	Ref.
Directed graphs (digraphs)	FTS (Fault tree synthesis)	3, 4
Decision tables	CAT (Computer automated tree)	5, 6
Mini-fault trees (based on functional equations)	FPW (Fault propagation work)	7
Mini-fault trees (based on mathematical equations)	RIKKE	8, 9

III. FAULT TREE SYNTHESIS BY COMPONENT MODELING

The first step in fault tree synthesis is the decomposition of the process plant into its constituent parts. The usual procedure is to view process plants as a set of equipment (such as pumps, pipes, and vessels) linked in accordance with the process flow diagram. This approach is called component modeling.[2] Several important methods associated with this approach are summarized in Table 1.

Component models are a set of logic/mathematical expressions describing the behavior of the component under both fault and working conditions. A component model can be presented in various forms, each associated with a particular method. The forms given in Table 1 are described briefly here.

The method of Lapp and Powers[3,4] uses a directed graph (digraph) to represent component models. The nodes in this digraph represent process variable deviations or component failures. The edges, which link the nodes, are weighted and correlated by a positive or negative sign. The magnitude of the weight implies the intensity of the gain between the two nodes and the sign indicates the direction of the deviation. A link may be conditioned by a failure.

The method of Salem et al.[6] employs decision tables as the form of component models. A decision table is an extended truth table with multiple inputs and outputs. Decision tables representing component models may consist of three main columns: component input variables, component failure states, and component output variables. Each row of the table is an alternative combination of the input variables and the internal states to result in a specific output variable.

Mini fault trees are the form adopted in the two methods of Martin-Solis et al.[7] (FPW) and Taylor[8] (RIKKE). This is the same concept that was first introduced by Fussell as the Failure Transfer Functions.[10] A component normally needs several sets of mini trees to define it. Each set of mini trees represents one mode of the component failure.

FPW (Fault Propagation Work) uses functional equations as a means for development of mini fault trees. A functional equation is a relation between the deviation of the component output and a number of the input deviations. The effect of an input on the output is indicated by the sign in front of the input. Component failure states are added to the model based on engineering judgment.

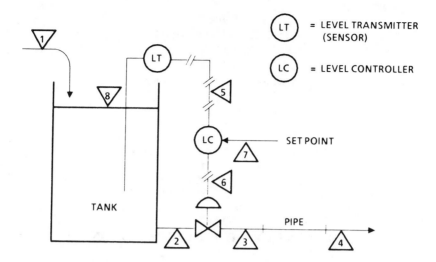

FIGURE 3. Process flow diagram of a level control system.

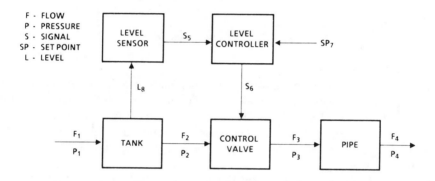

FIGURE 4. Block diagram of the level control system.

Instead of functional equations, RIKKE uses a perturbed form of mathematical equations. The deviation of the component output is no longer given by the input deviations separately. Hence the model is more complicated and at the same time more representative, because conditions are included in models.

Once component models are generated, the method assembles them according to the plant configuration under study. For example, consider Figure 3 showing a process flow diagram of a simple level control system. This flow diagram should be converted to the system configuration which consists information concerning the types and interconnections of the components in the system, Figure 4. When the component models have been assembled, the system is ready to develop fault trees for deviations of all process variables, or undesirable events associated with the variables.

IV. FAULT TREE SYNTHESIS BY STRUCTURE MODELING

In contrast to component modeling, structure modeling decomposes the system into a set of functional subsystems. Typically, these subsystems are the control loops. The objective is to define the failure behavior of the plant in terms of the control loops.

First, the control loop is analyzed to identify its main features such as its functions and its interrelations with other subsystems and with the environment. Then, attempts are made to show how the control loop structure fails.

Table 2
THE CONTROL LOOP
ELEMENTS

Main parts	Elements
Control instrument	Sensor
	Controller
	Control valve
Process	Other control loop
	hardware, e.g.,
	heat exchangers,
	valves, pumps, etc.

Table 3
THE CONTROL LOOP CLASSIFICATION

Class	Examples
Single loop	Feedback, feedforward, trip
Multiple loop	Cascade, feedback-feedforward
Manual control	Human operator
Self-regulating	Pressure relief valves

A. Control Loop Analysis

Physically, the control loop consists of two main parts: control instrument and process. Table 2 presents these parts with their elements. For the purpose of this book, there are four classes of control loop. These classes, with examples, are outlined in Table 3.

The function of the control loop is to maintain the controlled variable at the desired value or set point. Failure of the control loop is defined as deviation of the controlled variable from the desired limit.

The control loop has a single output, i.e., the controlled variable. It has, however, a number of inputs which are potentially able to perturb the controlled variable. Furthermore, there can be other inputs that pass through the loop uninfluenced. For example, in a control loop where the controlled variable is temperature, changes in the concentration of a substance can pass through uninfluenced.

The plant system consists of a number of control loops. The inputs to the loop are either inputs across the plant system boundary or outputs from other loops. The output from the loop is either an output across the plant system boundary or an input to other loops.

The environment encompasses everything outside the process plant system. The plant system is exposed to environmental effects which can perturb the controlled variable. Some elements of the environment are (1) utilities (steam, instrument air, and electricity); (2) climate; and (3) sometimes the process operator.

B. Control Loop Failure

Control loop failure occurs when controlled variable deviations occur. The cause for this deviation is called a disturbance. The various ways in which control loop failure can occur are identified by analyzing the disturbances.

Disturbances can be external and internal. An external disturbance is the result of a change in process variables input to the loop. The three types of external disturbances are shown in Table 4. The first type is a disturbance which saturates the control action available, so that although the valve is fully open or fully shut, there is still a controlled-variable deviation. The second type is a disturbance which lies within the range of potential correction of the

Table 4
EXTERNAL DISTURBANCE TYPES

External disturbance	Designation	Complementary condition
Uncontrollable disturbance	UD	—
Controllable disturbance	CD	Element invariant failure (SK)
Failure-inducing disturbance	FD	Element suscepti- bility failure (FS)

Table 5
INTERNAL DISTURBANCE TYPES

Internal disturbance	Designation	Complementary condition
Control or process element failure	EF	—
Control element invariant failure	SK	Controllable disturb- ance (CD)
Element susceptibility	ES	Failure-inducing dis- turbance (FD)

control but still causes a controlled-variable deviation because an element in the loop has failed in a fixed position or failed stuck. The third type is a disturbance which induces failure of a susceptible element in the loop.

An internal disturbance is the outcome of an equipment failure within the control loop boundary. The three types of internal disturbance are shown in Table 5. The first type is the case where the equipment has an inherent weakness and fails spontaneously. Control elements or process elements can fail catastrophically, and each can perturb the controlled variable. A control element, in the second type, can fail invariably. This failure cannot perturb the controlled variable unless a controllable disturbance enters the loop. The final type of internal failure is the case where the element has the susceptibility to failure and fails due to an external disturbance.

The disturbances described can lead to control loop failure. The different failure modes of the control loop are shown in Figure 5. This figure is called the control loop failure structure, which is often presented in tabular form.

In the case where it has protective device, the control loop fails when the controlled-variable deviation occurs and the device fails to protect. This is depicted in Figure 6.

V. METHOD OF FAULT TREE SYNTHESIS

In this section, an algorithm is presented for development of fault trees based on structure modeling. The main objective of the method is to construct fault trees in which the functional relations are more explicit and transparent than in fault trees produced by component modeling.

The process of fault tree synthesis consists of two main steps. In the first step, the generalized fault tree based on the plant digraph is constructed. The generalized fault tree is a hierarchy in which the top element is the control loop whose failure is the top event of the fault tree. The elements of the lower levels are identified based on the ways they are linked to the top control loop.

Consider Figure 7, which is the plant digraph of a hypothetical plant consisting of five control loops. In this digraph, the nodes represent the control loops, the edges are the controlled variables, and the disturbances come across the plant boundary.

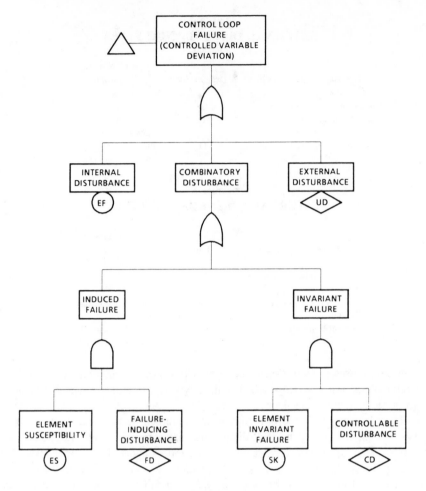

FIGURE 5.   Control loop failure structure.

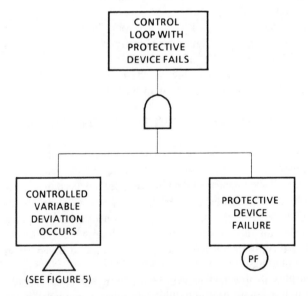

FIGURE 6.   Failure of control loop with protective device.

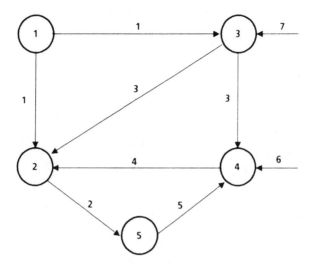

FIGURE 7. Plant digraph of a hypothetical plant.

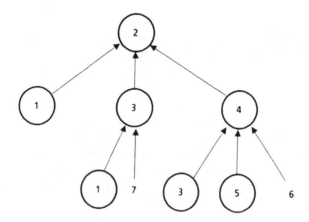

FIGURE 8. Generalized fault tree for control loop 2.

Suppose we are interested in the failure of Control Loop 2. The generalized fault tree of this control loop failure is shown in Figure 8. This figure shows how faults can propagate through the plant to result in the failure of Control Loop 2. A similar generalized fault tree can be developed for other control loops. The development of the hierarchy given in Figure 8 from the network shown in Figure 7 is not complicated, and an algorithm for the conversion has been described elsewhere.[11]

The second main step is to develop a specific fault tree based on the generalized fault tree. For this, we need to have the failure structures of the control loops. The specific fault tree is constructed by replacing each circle in the generalized fault tree by the appropriate control loop failure structure. The method of fault tree synthesis described is shown by means of a diagram in Figure 9.

A. Example of Fault Tree Synthesis

To illustrate this method of fault tree synthesis, consider the heating system of a propylene pipeline.[12] The purpose of this system is to raise the temperature of the propylene before it is transferred through a 7-mile long pipeline. The pipeline is made of mild steel, so that if

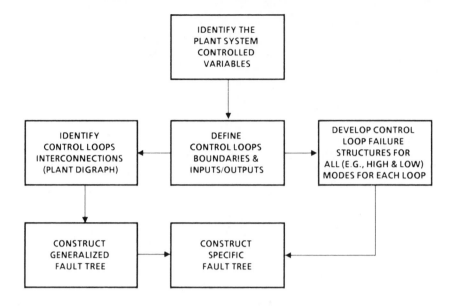

FIGURE 9.   Method of fault tree synthesis based on control loop structure.

the heating system fails to raise the temperature sufficiently, the pipeline will likely become brittle and may fracture. The propylene is heated by a heat transfer oil, and the oil is in turn heated by steam. The process flow diagram of the system is shown in Figure 10.

In this plant system, there are four controlled variables: (1) Propylene temperature, (2) Heat transfer oil temperature, (3) Heat transfer oil flow, and (4) Condensate level. Therefore, the plant is decomposed into four control loops.

The physical boundaries of the control loops are shown in Figure 10. Each loop is identified by a number; for example, the propylene temperature control loop is number 1. The disturbances entering the plant across the plant boundary are also numbered. For example, in Figure 10, the propylene flow is identified by number 5. Table 6 summarizes the control loops and the outside disturbances identification.

Among the control loops, Loop 3 is manually controlled and the rest are single-feedback loops. Loop 1 is the only one which has a protective device. The device activates "on" when a low propylene temperature occurs. The most serious event is when the temperature drops low enough to cause pipeline fracture and subsequent release of the material.

The analyses of the four loops are given in Table 7A to 10A. For each loop, the analysis involves the definition of the loop elements and the identification of the external and internal disturbances. The failure structures of the control loop are given in Tables 7B to 10B. Each table contains two different failure modes (HI and LO) of the controlled variable deviation.

The plant digraph interconnecting the controlled variables of the loops is shown in Figure 11. Links 2 to 4 correspond to the outputs of the corresponding control loops, and Links 5 to 8 are the four disturbances entering the plant across the boundary.

The fault tree is synthesized in two steps, but before these the top event (most serious event) must be chosen. As noted, the most serious event is the possible fracture of the mild steel pipeline. This is a deviation of the output from Loop 1 and the failure on demand of the protective system. However, the first step is the construction of the generalized fault tree. The tree is developed based on the plant digraph and is shown in Figure 12. In the second step, the specific fault tree is obtained from the generalized fault tree by replacing each circle by the appropriate failure structure given in Tables 7B to 10B. The specific fault tree for the top event of the pipeline fracture is shown in Figure 13.

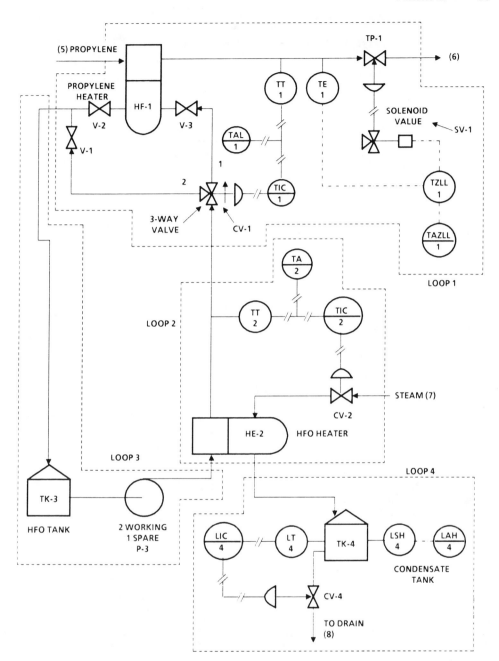

FIGURE 10. Process flow diagram of the propylene pipeline heating system.

VI. DISCUSSION

The complexity, the nature of materials used, and the operating conditions all contribute to the fact that process plants are regarded as sources of hazards. A hazard is potentially capable of inflicting risk on the employees and the public. Identifying hazards, estimating the associated risks, and controlling these risks are three main tasks of process system safety and reliability analysis. Essential in all these tasks is being systematic to be not only efficient, but also effective.

Table 6
CONTROL LOOPS/DISTURBANCE IDENTIFICATION

Control loop (C)/outside disturbance (D) number	Process variable
1 (C)	Propylene temperature
2 (C)	Oil temperature
3 (C)	Oil flow
4 (C)	Condensate level
5 (D)	Propylene flow
6 (D)	Propylene pipeline pressure
7 (D)	Steam supply
8 (D)	Condensate backup pressure

Table 7A
CONTROL LOOP 1 ANALYSIS

Loop Elements

Control Instrument (IN1)	Sensor 1 (TT-1); controller 1 (TIC-1) three-way valve 1 (CV-1) (Port 1 opens on air failure)
Process equipment	Heat exchanger (HE-1); by-pass valve (V1); block valves (V2 and 3)
Protective devices	Temperature trip sensor (TE-1); trip switch and solenoid valve (SV-1); trip valve (TP-1) (normally open)

External Disturbances

Disturbance	Type	Complementary condition	Controlled variabled deviation
Q5 HI	UD	—	T1 LO
Q5 LO	CD	IN1 SK	T1 HI
P6 HI	CD	IN1 SK	T1 HI
P6 LO	UD	—	T1 LO
Q3 HI	CD	IN1 SK	T1 HI
Q3 LO	UD	—	T1 LO
T2 HI	CD	IN1 SK	T1 HI
T2 LO	UD	—	T1 LO
SP-1[a] HI	UD	—	T1 HI
SP-1 LO	UD	—	T1 LO

Internal Disturbances

Disturbance	Type	Complementary condition	Controlled variabled deviation
HE-1 FOULING	EF	—	T1 LO
V-1 CLOSED	EF	—	T1 HI
V-2/V-3 CLOSED	EF	—	T1 LO
TT-1 F HI	EF	—	T1 LO
TT-1 F LO	EF	—	T1 HI
TIC-1 F HI	EF	—	T1 LO
TIC-1 F LO	EF	—	T1 HI
TIC-1 FAILED	ES	IN. AIR LO	T1 HI
CV-1, Port 1 F CLOSED	EF	—	T1 LO
CV-1, Port 2 F CLOSED	EF	—	T1 HI

[a] SP; Set Point.

Table 7B
CONTROL LOOP 1 FAILURE STRUCTURES

Failure structure T1 HI

Q5 LO AND IN1 SK
P6 HI AND IN1 SK
Q3 HI AND IN1 SK
T2 HI AND IN1 SK
V-1 CLOSED
TT-1 F LO
TIC-1 F LO
TIC-1 FAILED AND IN. AIR LO
CV-1, Port 2 CLOSED
SP-1 H1

Failure structure T1 LO

Q5 HI
P6 LO
Q3 LO
T2 LO
HE-1 FOULING
V-2/V-3 CLOSED
TT-1 F HI
TIC-1 F HI
CV-1, Port 1 CLOSED
SP-1 LO

Complete control system failure

Since T2 LO is a trip condition and the Loop has a protective device, the whole control system fails under the following condition:

T1 LO AND protective device failure (PF)

PF = TE-1 F on D[a], or
SV-1 F on D, or
TP-1 F on D.

[a]   F on D; fails on demand.

Fault tree analysis is only one important part of the second task of system safety analysis in which the risk is estimated. In order to be systematic, attempts were made in the past to synthesize fault trees. Most works were based on the concept of component models, components being equipment items.

A method described in this paper is based on structure modeling. As the result of this method, fault trees have a more clearly defined structure and are more transparent. The main effort required in the application of the method is the development of the control loop analysis and failure structures. The efforts made to decompose the process plant and analyze the control loops give the analyst valuable insights into the system behavior.

In the example used in this chapter, several system failure contributors, such as human errors, were not considered rigorously. Nevertheless, the role of human operator is in fact defined clearly in this method. That is, where a loop is considered to be manually controlled, the human operator is viewed as the controller. Therefore, specific duties are expected of the operator. In other cases, the operator's roles are mainly related to the indicators, the alarms, and the set points.

Table 8A
CONTROL LOOP 2 ANALYSIS

Loop Elements

Control instrument (IN2)	Sensor 2 (TT-2); controller 2 (TIC-2); control valve 2 (CV-2) (air-to-close)
Process equipment	Heater 2 (HE-2)

External Disturbances

Disturbance	Type	Complementary condition	Controlled variable deviation
Q3 HI	UD	—	T2 LO
Q3 LO	CD	IN2 SK	T2 HI
L4 HI	UD	—	T2 LO
L4 HI	—	—	None
Q7 HI	CD	IN2 SK	T2 HI
Q7 LO	UD	—	T2 LO
SP-2 HI	UD	—	T2 HI
SP-2 LO	UD	—	T2 LO

Internal Disturbances

HE-2 FOULING	EF	—	T2 LO
TT-2 F HI	EF	—	T2 LO
TT-2 F LO	EF	—	T2 HI
TIC-2 F HI	EF	—	T2 LO
TIC-2 F LO	EF	—	T2 HI
TIC-2 FAILED	ES	IN. AIR LO	T2 HI
CV-2 F OPEN	EF	—	T2 HI
CV-2 F CLOSED	EF	—	T2 LO

Table 8B
CONTROL LOOP 2 FAILURE STRUCTURES

Failure structure T2 HI

Q3 LO AND IN2 SK
Q7 HI AND IN2 SK
TT-2 F LO
TIC-2 F LO
TIC-2 FAILED AND IN. AIR LO
CV-2 F OPEN
SP-2 HI

Failure structure T2 LO

Q3 HI
L4 HI
Q7 LO
HE-2 FOULING
TT-2 F HI
TIC-2 F HI
CV-2 F CLOSED
SP-2 LO

Table 9A
CONTROL LOOP 3 ANALYSIS

Loop elements

Control instrument (IN3)	Human operator
Process equipment	HFO Tank (TK-e); Pumps (P-3); Pipework (PP-3)

Internal disturbances

Disturbance	Type	Complementary condition	Controlled variabled deviation
P-3 F to R[a]	EF	—	Q3 LO
PP-3 LEAK	EF	—	Q3 LO
PP-3 BLOCK	EF	—	Q3 LO
TK-3 LEAK	EF	—	Q3 LO

[a] F to R; fails to run.

Table 9B
CONTROL LOOP 3 FAILURE STRUCTURES

Failure structure Q3 LO

P-3 F to R[a]
PP-3 LEAK
PP-3 BLOCK
TK-3 LEAK

[a] F to R; fails to run.

Another main contributor is the failure which results in multiple failures in the system. The environment, the human operator, and the utilities are the obvious causes of multiple failures. The classification of the control loop internal failures helps define a component failure as "element susceptibility", to provide a way to include a cause of multiple failures.

Table 10A
CONTROL LOOP 4 ANALYSIS

Loop Elements

Control instrument (IN4)	Sensor 4 (LT-4);
	controller 4 (LIC-4);
	control valve (CV-4) (air-to-open)
Process Equipment	Condensate tank (TK-4)

External Disturbances

Disturbance	Type	Complementary condition	Controlled variable deviation
P8 HI	UD	—	L4 HI
P8 LO	—	—	None
SP-4 HI	UD	—	L4 HI
SP-4 LO	UD	—	L4 LO

Internal Disturbances

Disturbance	Type	Complementary condition	Controlled variable deviation
LT-4 FHI	EF	—	L4 LO
LT-4 FLO	EF	—	L4 HI
LIC-4 FHI	EF	—	L4 LO
LIC-4 FLO	EF	—	L4 HI
LIC-4 FAILED	ES	IN. AIR LO	L4 HI
CV-5 F OPEN	EF	—	L4 LO
CV-5 F CLOSED	EF	—	L4 HI

Table 10B
CONTROL LOOP 4 FAILURE STRUCTURES

Failure structures L4 HI

P8 HI
LT-4 F LO
LIC-4 F LO
LIC-4 FAILED AND IN. AIR LO
CV-4 F CLOSED
SP-4 HI

Failure structure L4 LO

LT-4 F HI
LIC-4 F HI
CV-4 F OPEN
SP-4 LO

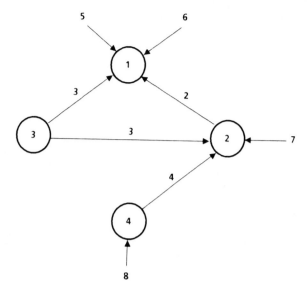

FIGURE 11. Plant digraph for the propylene heating system.

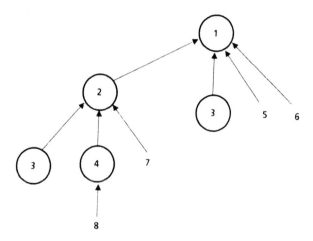

FIGURE 12. The generalized fault tree for the control loop 1 failure.

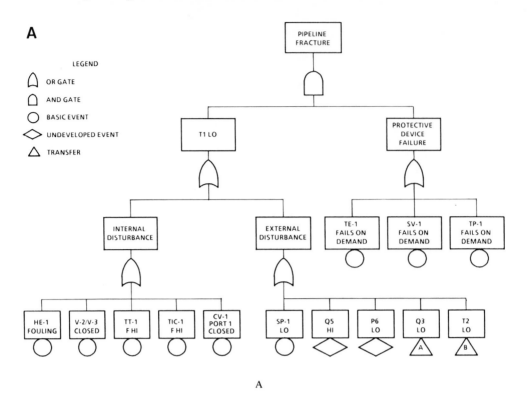

A

FIGURE 13.   Specific fault tree.

B

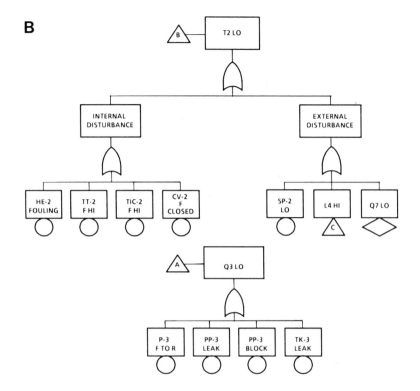

FIGURE 13B.

C

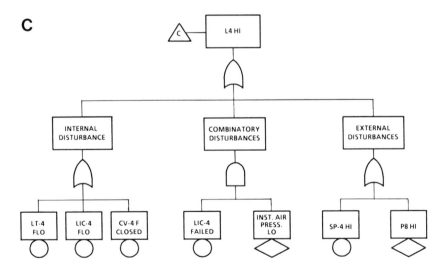

FIGURE 13C.

REFERENCES

1. **Shafaghi, A. and Gibson, S. B.,** Hazard and Operability (HAZOP) Study: a Flexible Technique for Process System Safety and Reliability Analysis, presented at the 187th American Chemical Society National Meeting, St. Louis, Missouri, April 8 to 13, 1984.
2. **Shafaghi, A.,** Component Modeling for Fault-Tree Synthesis, Interim Report, Battelle Columbus Division, Columbus, Ohio, 1982.
3. **Lapp, S. A. and Powers, G. J.,** Computer-aided synthesis of fault trees, *IEEE Trans. Reliab.,* R-26(1), 2, 1977.
4. **Lapp, S. A. and Powers, G. J.,** Update of Lapp-Powers fault-tree synthesis algorithm, *IEEE Trans. Reliab.,* R-28(1), 12, 1979.
5. **Apostolakis, G. F., Salem, S. L., and Wu, J. S.,** CAT: A Computer Code for the Automated Construction of Fault Trees, EPRI NP-705, Electric Power Research Institute, Palo Alto, Calif., 1978.
6. **Salem, S. L., Apostolakis, G. E., and Okrent, D.,** A Computer-Oriented Approach to Fault-Tree Construction, EPRI NP-288, Electric Power Research Institute, Palo Alto, Calif., 1976.
7. **Martin-Solis, G. A., Andow, P. K., and Lees, F. P.,** Fault-tree synthesis for design and real time applications, *Trans. Inst. Chem. Eng.,* 60(1), 14-25, 1982.
8. **Taylor, J. R.,** *A Background to Risk Analysis,* Vols. 1-4, Riso National Lab., DK-4000, Roskilde, Denmark, 1979.
9. **Taylor, J. R.,** An algorithm for fault-tree construction, *IEEE Trans. Reliab.,* R-31(2), 137, 1982.
10. **Fussell, J. B.,** A formal methodology for fault-tree construction, *Nucl. Sci. Eng.,* 52, 421, 1973.
11. **Shafaghi, A., Andow, P. K., and Lees, F. P.,** Fault-tree synthesis based on control loop structure, *Chem. Eng. Res. Des.,* 62, 101, 1984.
12. **Kletz, T. A.,** Practical applications of hazard analysis, *Chem. Eng. Prog.,* 34, 1978.

Chapter 4

RISK ANALYSIS OF FATIGUE FAILURE OF HIGHWAY BRIDGES

Nur Yazdani and Pedro Albrecht

TABLE OF CONTENTS

I. INTRODUCTION

Over one half of the approximately 600,000 bridges in the U.S. are more than 30 years old, with a large number of them needing rehabilitation. On the average, 12,000 bridges reach their design life every year. Since there are not enough funds to continuously replace bridges at that rate, one has no choice but to keep in service beyond the design life, for as long as possible, those bridges which have not become structurally deficient. There is at present no methodology available to help the bridge engineer assess the risk of a fatigue/fracture failure if the service life is extended. Such a tool can assist the engineer in determining when to replace or rehabilitate a bridge.

The objective of this study was to develop a model for calculating the risk of fatigue failure of a steel bridge. This required a probabilistic fracture mechanics calculation of crack growth with stochastic inputs for crack growth rate, fracture toughness, crack size, and load history. The model was checked against the known service failure of a highway bridge.

This chapter summarizes the work reported in detail in Reference 12.

II. DETERMINISTIC MODEL

The basic deterministic model used herein to describe the crack growth rate is given by the Paris equation:

$$\frac{da}{dN} = C(\Delta K)^n \tag{1}$$

in which a = crack size, N = number of cycles, C and n = material constants, and ΔK = range of stress intensity factor. Equation 1 describes the crack growth behavior in the intermediate region, II, of the crack growth rate curve shown in Figure 1. It was extended on the lower end below the threshold value of the stress intensity factor range, ΔK_{th}, and cut off on the high end where the stress intensity factor at the peak of a load cycle equaled the fracture toughness, K_{IC}.

The stress intensity factor range was expressed as:[2]

$$\Delta K = F_e F_s F_w F_g \, \sigma_r \sqrt{\pi a} \tag{2}$$

in which σ_r = stress range; and F_e, F_s, F_w, and F_g = crack shape, free surface, finite width, and stress gradient correction factors, respectively. The effect of the stress ratio, $R = \sigma_{min}/\sigma_{max}$, was accounted for by modifying Equation 1:[10]

$$\frac{da}{dN} = C\left(\frac{\Delta K}{1 - R/Q}\right)^n = C(\Delta K_{eff})^n \tag{3}$$

The term Q was determined by analysis of the crack growth data,[14] and $\Delta K_{eff} = \Delta K/(1 - R/Q)$.

The fatigue life of a structural detail was obtained by numerical integration of Equation 3 from the initial crack size, a_i, to the final crack size, a_f:

$$N = \int_{a_i}^{a_f} \frac{da}{C\left(\dfrac{\Delta K}{1 - R/Q}\right)^n} \tag{4}$$

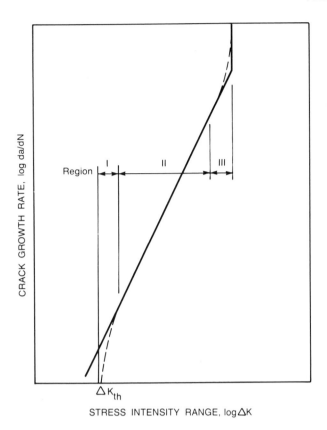

FIGURE 1. Regions of fatigue crack growth. (From Yazdani, N., *J. Struct. Eng.*, 113(3), 487, 1987. With permission.)

Fatigue cracks initiate from flaws introduced during rolling, fabrication, and handling. They grow in three stages: (1) part-through crack, (2) two-ended crack, and (3) three-ended crack. Figure 2 illustrates, as an example, the crack growth at a transverse stiffener welded to the web of a steel beam.[5] The first stage consumed most of the fatigue life and is, therefore, the most important. However, all three stages were modeled.

The free surface and the finite width corrections on the stress intensity factor, F_s and F_w, depend on the stress gradient and the semi-axis ratio of the elliptical crack. In this study, independent correction factors were assumed for the Category A and B details, which do not have a stress gradient. Dependent correction factors were used for Category C and E details in which the weld toe creates a steep stress gradient.[13,15]

The final crack size, a_f, in each stage of crack growth was calculated by substituting the value of fracture toughness, K_{Ic}, for the stress intensity factor, K, in Equation 2, and solving it numerically for crack size.

III. CRACK GROWN RATE DATA

The material constants C and n determine the rate of crack growth in a given steel and environment. Data for crack growth rates in mild, high-strength low-allow (HSLA) and quenched and tempered (Q&T) structural steels exposed to air and aqueous environments were collected from the literature. Over 3,500 points were digitized and regression analyses were performed. The data points falling in the transition regions, I and III, (Figure 1) were excluded from the analysis.

Stage 1: Part - Through Crack
Stage 2: Two - Ended Through Crack
Stage 3: Three - Ended Crack

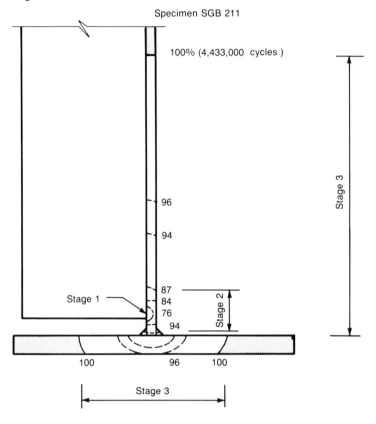

FIGURE 2.   Stages of crack growth at web stiffeners. Stage 1: part-through crack.
Stage 2: two-ended through crack. Stage 3: three-ended crack. (From Yazdani, N.,
J. Struct. Eng., 113(3), 487, 1987. With permission.)

The statistical difference between any two mean lines was checked with the T-test at the
5% level of significance. Significant differences were found between crack growth rates in
air vs. aqueous environments, and between mild and HSLA versus Q&T steels. As a result,
the following crack growth equations were obtained. Mild and HSLA steels in air:

$$\frac{da}{dN} = 8.291 \times 10^{-11}(\Delta K_{eff})^{3.344} \tag{5}$$

Mild and HSLA steels in aqueous environments:

$$\frac{da}{dN} = 2.23 \times 10^{-10}(\Delta K_{eff})^{3.279} \tag{6}$$

Quenched and tempered steels in air:

$$\frac{da}{dN} = 1.174 \times 10^{-9}(\Delta K_{eff})^{2.534} \tag{7}$$

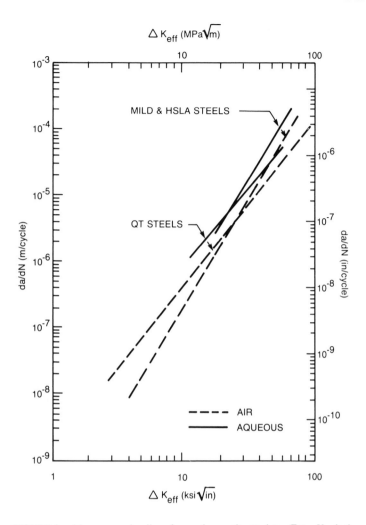

FIGURE 3.   Mean regression lines for crack growth rate data. (From Yazdani, N., *J. Struct. Eng.*, 113(3), 487, 1987. With permission.)

Quenched and tempered steels in aqueous environments:

$$\frac{da}{dN} = 2.975 \times 10^{-9}(\Delta K_{eff})^{2.420} \tag{8}$$

In the above equations, da/dN and ΔK_{eff} have units of in/cycle and ksi$\sqrt{in}$, respectively. The four equations were plotted in Figure 3. The air and aqueous lines are parallel for each group of steel, with the cracks growing about twice as fast in aqueous than in air environments.

IV. FRACTURE TOUGHNESS DATA

The fracture toughness depends on the type of steel, temperature, and loading rate. A comprehensive collection of plane-strain fracture toughness (K_{Ic}) and Charpy V-notch impact (CVI) data for structural steels was taken as the data base for this study.[9] Three types of loading were considered: (1) static loading lasting at least 100 sec, (2) 1-sec loading, and (3) 1-msec dynamic loading. The 1-sec loading closely approximates the loading rate of short-span bridges.

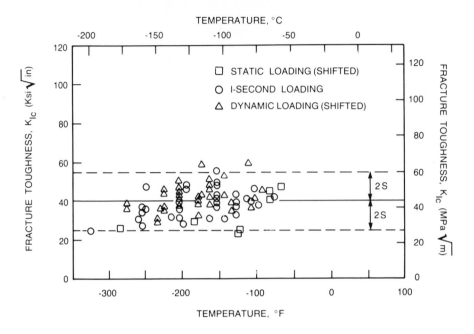

FIGURE 4.   Fracture toughness data for A36 steels. (From Yazdani, N., *J. Struct. Eng.*, 113(3), 487, 1987. With permission.)

The CVI data were converted to equivalent dynamic fracture toughness data. All static and dynamic data were then converted to equivalent 1-sec loading data by using a temperature shift. The temperature shift between the static and dynamic fracture toughness curves is given by:[9]

$$T = 215 - 1.5 F_y \qquad (9)$$

in which T is the shift in °F and F_y is the yield strength of the steel. The shift between the 1-sec and dynamic loadings was found to be three fourths of T. The data for A36 steels are shown in Figure 4.

The distribution of the fracture toughness data was checked with the Chi-square goodness-of-fit test at $\alpha = 5\%$. The normal distribution best fitted the fracture toughness data for all three steels. The lower tail of the distribution was truncated at zero fracture toughness.

V. CRACK SIZE DATA

The initial crack size distribution means the conditional distribution, given that a crack is initially present. There is no information available on initial crack sizes at structural details. Therefore, they had to be calculated from stress range vs. cycles to failure (S-N) data. [5-7] Because these data are the basis for the AASHTO allowable S-N lines, they should also be acceptable for determining equivalent initial crack sizes. The initial crack size for each type of detail was calculated by backward integration of Equation 4 over the known fatigue life, beginning at the final crack.

The results of the Chi-square goodness-of-fit test at $\alpha = 5\%$ showed that most data sets for initial crack size were log-normally distributed. For reasons of uniformity, the distribution of all data sets were assumed to be log normal. Figure 5 shows a typical histogram for the initial size (depth) of part-through cracks at cover plate ends.

The bridges analyzed in this study were assumed to be periodically inspected by the

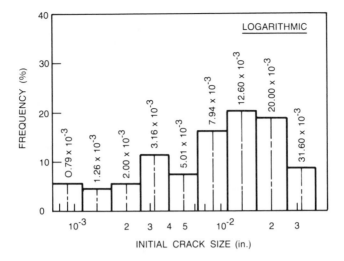

FIGURE 5. Frequency histograms of initial crack sizes for mild and HSLA cover plates. (From Yazdani, N., *J. Struct. Eng.*, 113(3), 487, 1987. With permission.)

ultrasonic method. The following log-normal model describes the probability of not detecting a part-through crack with ultrasonics:[8]

$$P_{ND} = 0.5 \text{ erfc}\left(1.46 \ln \frac{a}{0.35}\right) \tag{10}$$

in which a = crack size in inches. Equation 10 can be changed to represent the data for any other method of nondestructive inspection.

VI. LOAD HISTORY DATA

The load history of a bridge is given by the stress range induced by trucks crossing the bridge and by the average daily truck traffic (ADTT). A total of 162 stress range histograms were collected from reported strain gage data that had been recorded on 40 bridges in 9 states.[12] Because these data were recorded between 1963 and 1973, they may be a nonconservative estimate of present day truck loading.

Each stress range histogram was replaced by an equivalent constant amplitude stress range. The equivalent stress range was then normalized with respect to the design stress range. The Chi-square goodness-of-fit test at $\alpha = 5\%$ was performed on the two largest data sets of normalized equivalent stress ranges that were found to be log-normally distributed, those for Category B details on Illinois bridges and Category A details on Ohio bridges. The other data sets were too small for meaningful statistical tests.

The ADTT was calculated as follows:

$$\text{ADTT} = \frac{(\text{No. of trucks counted})}{(\text{No. of recording hours})} (24 \text{ hours}) \tag{11}$$

The Alabama and Maryland bridges had the lowest ADTT of the instrumented bridges in the nine states; the Ohio and Connecticut bridges had the highest ADTT.

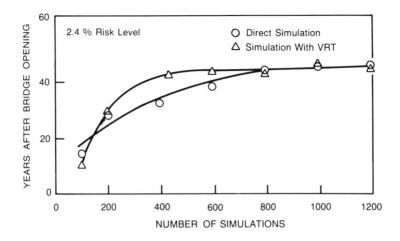

FIGURE 6.   Number of simulations needed to achieve a 5% confidence interval. (From Yazdani, N., *J. Struct. Eng.*, 113(3), 487, 1987. With permission.)

VII. PROBABILISTIC MODEL

The probabilistic model consisted of the stochastic input data, deterministic fracture mechanics calculation of crack growth, and Monte Carlo simulation with variance reduction to obtain the distribution of the output variable. The four independent input variables were the crack growth rate, fracture toughness, initial crack size, and equivalent stress range. The truncated probability density function (pdf) is defined by:

$$f(x) = \frac{g(x)}{\int_a^b g(x)\, dx} \tag{12}$$

in which $g(x)$ = pdf for input variable, $a < x < b$, and a and b = lower and upper truncation limits, as applicable. The cumulative distribution functions (cdf) were calculated from individual pdfs by numerical integration.

The pseudo-random numbers needed as initial seeds for generating the four input variables were the instantaneous hours, minutes, seconds, and date from the computer clock. The stimulated input variable was interpolated from the corresponding cdf array by the inverse-function method.

Variance reduction techniques (VRT) reduce the required number of simulations when the risk of failure is small. Conversely, for an equal number of simulations, the variance of failure risk is smaller than that obtained by direct simulation. Using the antithetic VRT,[3] it was possible to reduce from 800 to 500 the required number of simulations needed to obtain a 5% confidence interval, as seen in Figure 6.

Most highway bridges are redundant load-path (RLP) structures, meaning that the bridge does not collapse when a detail fails. However, as soon as a crack is detected, the common practice is to close the bridge for inspection and repair. A RLP bridge can, therefore, be viewed as being nonredundant in terms of the number of traffic paths. A nonredundant load path (NRLP) bridge fails when one detail fails. It is structurally nonredundant.

In the sense that both RLP and NRLP bridges are taken out of service when a detail fails, until the bridge is repaired or rebuilt, respectively, both can be viewed as being nonredundant in terms of traffic path and use. The risk of failure of the bridge (system in series) is, therefore, bounded by:

$$\max P_{F_i} \leq P_{F_s} \leq 1 - \sum_{i=1}^{k} (1 - P_{F_i}) \tag{13}$$

in which P_F = system risk, P_F = detail risk, and k = number of details in the system. The lower and upper bounds in Equation 13 correspond to correlation coefficients between details of $\rho = 1$ and $\rho = 0$, respectively. The correlation between the failure probability of bridge details should be high; the exact values are difficult to estimate. An approximate correlation of $\rho = 0.7$ was assumed in this study. After calculating the bounds in Equation 13, the actual failure risk for $\rho = 0.7$ was determined by linear interpolation.

The system reliability depends on the number of details which make up the system (Equation 13). In a bridge, the system includes details on all girders falling longitudinally in a series. But the truck weight is mostly carried by two or three girders, depending on truck width and girder spacing. So, the total number of details in the system was calculated from:

$$k = \left[\left(\frac{\text{truck width}}{\text{girder spacing}} \right) \text{int} + 1 \right] gd \tag{14}$$

In which g = number of girders and d = number of details on each girder.

The values of the following variables can be specified when the model is used to calculate the cumulative risk as a function of time: (1) maximum length of service life extension, (2) specified inspection intervals, and (3) maximum failure risk at which the bridge must be inspected. The latter was assumed to be 2.4% for RLP structures, the same value as for the AASHTO allowable S-N lines. The acceptable risk of failure for NRLP structures should be much lower, because the consequences of failure are much higher in terms of loss of human life and cost of bridge replacement. The risks of fatigue failure at different phases of service life is as follows. Before the bridge is opened to traffic, or immediately after an in-service inspection:

$$P_F = P_{F_d} \tag{15}$$

in which P_{F_d} = design risk. Any time before the first or before a subsequent in-service inspection:

$$P_F = \frac{1}{N_s} \sum_{i=1}^{N_s} F_n P_{ND} + P_{f_d} \tag{16}$$

in which $F_n = 0$ for details that have not failed, $F_n = 1$ for details that have failed and N_s = number of simulations. P_{ND} is equal to 1.0 before the first in-service inspection, and equal to the value calculated with Equation 10 before a subsequent in-service inspection.

The design risk was claculated from:

$$P_{F_d} = \int_{t_f}^{\infty} a(x) \, dx \tag{17}$$

in which a(x) is the crack size distribution at any time during the service life, and t_f is the material thickness. It was assumed that the failed details would be repaired and their number subtracted from the number of details in the system for the subsequent simulations.

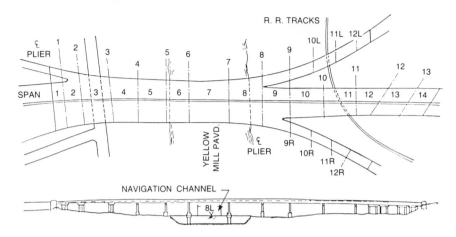

FIGURE 7.   Plan and elevation of Yellow Mill Pond Bridge. (From Yazdani, N., *J. Struct. Eng.*, 113(3), 487, 1987. With permission.)

VIII. RISK ANALYSIS OF SELECTED BRIDGES

The risk model developed herein was applied to three coverplated girder bridges built between 1958 and 1960.

Yellow Mill Pond Bridge, Connecticut — The Yellow Mill Pond Bridge on Interstate I-95 in Connecticut, built in 1958, consists of simple span steel girders with composite concrete deck (Figure 7). The A242 (HSLA) rolled steel girders have category E' cover plates.[7] In 1970, during the inspection of the repainting job done by a contractor, a large crack was discovered in span 11 of the eastbound roadway. The crack originated at the toe of the transverse fillet weld connecting the cover plate end to the tension flange of the beam, and extended up to 16 in. (400 mm) into the web. An inspection of 15 similar locations revealed additional cracks of varying magnitude along eight cover plate end welds. The two girders adjacent to the fractured girder had cracks extending halfway through the thickness of the tension flange.

In December 1970, after the bridge had been temporarily closed, the three damaged girders were repaired with bolted web and flange splices. During subsequent inspections in 1973, 1976, 1977, and 1979, fatigue cracks were detected at various other cover plate ends.[4] All girders with cracks that were at least $1 \frac{1}{2}$ in. (38 mm) long were repaired with bolted splices. During 1981, the remaining details were retrofitted by peening or bolting. The bridge had a high ADTT of 5,660 and a mean equivalent stress range of $f_{r,e} = 1.2$ ksi (29.7 MPa).[12]

The bridge consists of 14 consecutive simple spans with seven girders per span. For purposes of risk analysis, the system was assumed to consist of 14 spans with 3 girders (below the main truck lane) per span. Five girders had no cover plates, 8 had full-length cover plates, 21 had one cover plate and 8 had two cover plates. Using Equation 14, the number of details in the system was found to be 74.

Because the bridge was temporarily closed after the first set of cracks was discovered, it was considered to be nonredundant in terms of traffic paths. The bridge is load-path redundant.

Figure 8 presents the failure risk of one detail and the system, assuming the bridge is not inspected up to 80 years of service. As one would expect for nonredundant traffic-path structures, the system risk is much higher (about eight times at 50 years) than the detail risk. From the results for the system, one can state that 1.5% of bridges similar to the Yellow Mill Pond Bridge should fail after 12 years of service. In reality, the bridge was closed and repaired in the 12th year, when a girder had failed. The risk analysis program correctly predicts a very short service life.

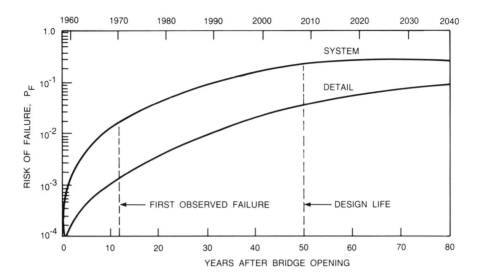

FIGURE 8. Comparison of detail and system reliability for Yellow Mill Pond Bridge, Connecticut; no inservice inspections. (From Yazdani, N., *J. Struct. Eng.*, 113(3), 487, 1987. With permission.)

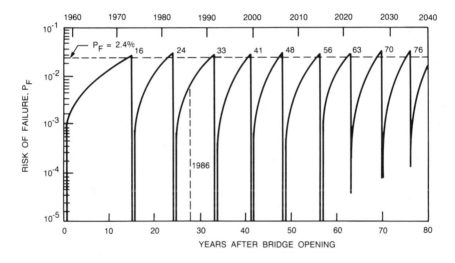

FIGURE 9. Effect of inservice inspection on failure risk of Yellow Mill Pond Bridge, Connecticut. ADTT = 5660 (1970; $f_{r,e}$ = 1.2 ksi (1970). (From Yazdani, N., *J. Struct. Eng.*, 113(3), 487, 1987. With permission.)

The allowable AASHTO S-N line for Category E' details predicts the safe service life (2.4% risk of failure) of the Yellow Mill Pond Bridge to be 99 years.[16] This value is eight times longer than the 12 years to first observed girder failure. The AASHTO fatigue specification is nonconservative, mainly because it accounts for the variability in the S-N data alone, as determined from one series of laboratory tests. It does not consider variabilities in crack growth rates, load history and fabrication.

Figure 9 presents the time intervals at which the bridge should be inspected to limit the risk to 2.4%. A total of nine inspections by the ultrasonic method, spaced from nine years at the beginning to six years towards the end, would be required in 80 years of service to limit the failure risk to 2.4%. In reality, the Yellow Mill Pond Bridge was inspected in 1970, 1973, 1976, 1977, and 1979.

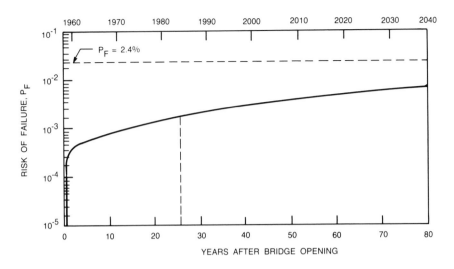

FIGURE 10.   Risk analysis of Marlboro Bypass Bridge, Maryland. ADTT = 522 (1970); $f_{r,e}$ = 1.33 ksi (1970). (From Yazdani, N., *J. Struct. Eng.*, 113(3), 487, 1987. With permission.)

Thorough inspections are time consuming and expensive. The intervals could be increased by reducing truck weight or traffic, or by preventive repair of details before they become fatigue critical. The first two options are not viable for an interstate highway. The third, retrofitting with a bolted splice, would fatigue-proof the cover plate end regardless of crack size.[1] But, this type of preventive repair is expensive. Other preventive repair methods, such as grinding, shot-peening or TIG remelting the weld toe, are less expensive, but only effective for shallow part-through cracks of about $1/_8$-in. (3 mm) depth.

A less expensive alternative to ultrasonic inspection at the calculated intervals would be to visually inspect the cover plate ends every year for through-thickness cracks that have not yet grown to a critical size. Through cracks break the paint coating, leaving visible rust stains. It is questionable, however, whether the low reliability of visual inspection would be generally acceptable.

Marlboro Bypass Bridge, Maryland — This bridge, built in 1960 on U.S. 301 near MD 4, consists of rolled girders with Category E cover plates. In 1970, the mean equivalent stress range was 1.33 ksi (9.2 MPa) and the ADTT was 522.[12] A 1982 traffic count by the State of Maryland showed that the traffic had risen about fourfold to ADTT = 2148. This bridge had about the same stress range history as the Yellow Mill Pond Bridge, but about 60% lower ADTT and details one AASHTO category higher. So, the early conditions for crack growth were much less severe than those in the Yellow Mill Pond Bridge. Indeed, no cracks were detected to date on this bridge. The bridge consists of three simple spans of seven girders each, with two cover plate ends per girder. The system was found to consist of 18 details.

Figure 10 presents the risk analysis results for 1970 estimates of $f_{r,e}$ and ADTT. It shows that no inspections would be needed for more than 80 years to keep the bridge open at that loading, without exceeding a 2.4% risk of failure.

Three parametric analyses were performed to determine what effect changes in loading would have on the risk of failure. First, the equilvalent truck weight (stress range) was increased by 10% and 20% to 1.46 and 1.60 ksi (10.1 and 11.0 MPa), respectively. Figure 11 shows that the latter case would advance the need for the first inspection to 72 years. Second, leaving the equivalent stress range at the 1970 value, but increasing the ADTT about fourfold over the entire life to the 1982 value of 2148 would require a first inspection after 77 years of service (Figure 12).

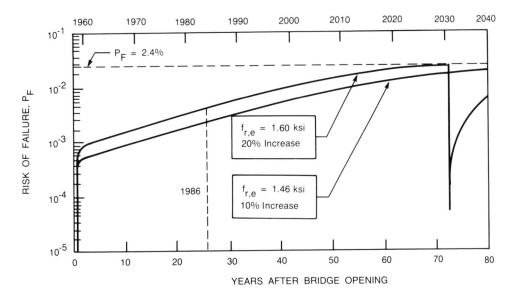

FIGURE 11.   Effect of 10% and 20% increase in truck weight on failure risk of Marlboro Bypass Bridge, Maryland; ADTT = 522 (1970). (From Yazdani, N., *J. Struct. Eng.*, 113(3), 487, 1987. With permission.)

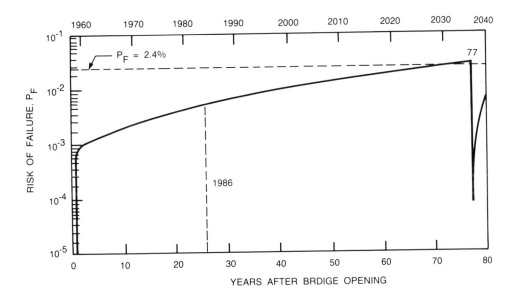

FIGURE 12.   Effect of increase in ADTT on failure risk of Marlboro Bypass Bridge, Maryland; ADTT = 2148 (1982), $f_{r,e}$ = 1.33 ksi (1970). (From Yazdani, N., *J. Struct. Eng.*, 113(3), 487, 1987. With permission.)

Third, the effect of a combined increase in $f_{r,e}$ to 1.60 ksi (11.0 MPa) and ADTT to 2148 is shown in Figure 13. This advances the first inspection to 54 years.

The time to first inspection increases rapidly with increasing truck weight and traffic. Subsequent inspection intervals are progressively shorter. Adequate safety built into the initial design can delay the need for a first thorough inspection for cracks until after the end of the 50-year design life, at a great savings in cost.

The Marlboro Bypass Bridge has a low risk of failure. The truck weight and traffic should be occasionally monitored.

Warrior River Bridge, Alabama — This bridge, located on U.S. 31 over the Warrior

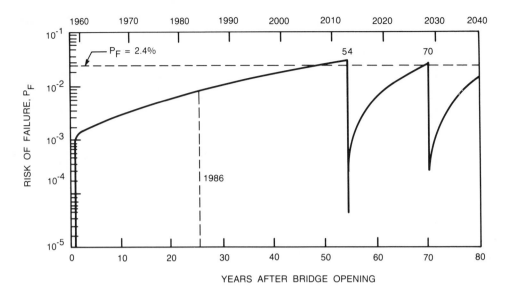

FIGURE 13.   Effect of increase in truck weight and ADTT on failure risk of Marlboro Bypass Bridge, Maryland; ADTT = 2148 (1982); $f_{r,e}$ = 1.60 ksi (20% increase). (From Yazdani, N., *J. Struct. Eng.*, 113(3), 487, 1987. With permission.)

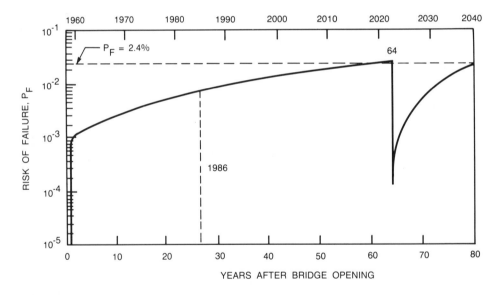

FIGURE 14.   Risk analysis of Warrior River Bridge, Alabama; ADTT = 1048 (1969), $f_{r,e}$ = 2.45 ksi (1969). (From Yazdani, N., *J. Struct. Eng.*, 113(3), 487, 1987. With permission.)

River in the Blount County, was opened to traffic in 1959. It consists of rolled girders with Category E cover plates. In 1969, $f_{r,e}$ was 2.45 ksi (16.9 MPa) and ADTT was 1048.[12] In 1984, a 15-min truck traffic count on U.S. 31 about 50 miles (80 km) away from the bridge site indicated an ADTT of 4128, about a fourfold increase. No cracks have been reported so far on this bridge. The bridge consists of three simple spans at five girders each, and two cover plate ends per girder. The system was found to consist of 18 details.

Figure 14 presents the risk analysis results with the 1969 values of $f_{r,e}$ and ADTT. One inspection would be required at 64 years to keep the failure risk at 2.4% or less, if $f_{r,e}$ and ADTT did not change.

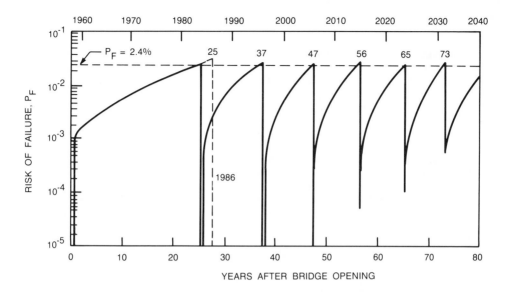

FIGURE 15. Effect of increase in ADTT failure risk of Warrior River Bridge, Alabama; ADTT = 4192 (1984), $f_{r,e}$ = 2.45 ksi (1969). (From Yazdani, N., *J. Struct. Eng.*, 113(3), 487, 1987. With permission.)

Because $f_{r,e}$ was already high in 1969 (about twice as high as on the Yellow Mill Pond bridge), in the parametric study only the ADTT was increased to its present day estimate of 4128. The results are shown in Figure 15. The increase in ADTT advances the first inspection to 1984, after 25 years of service. Five more inspections would be required in 80 years of service.

The recent opening of the Interstate highway I-65 is expected to reduce by 80% the truck traffic on U.S. 31 at the bridge site. This will drastically slow the increase in the risk of failure. However, considering that in 1984, the risk of failure was approaching 2.4%, and recalling the experiences with the Yellow Mill Pond Bridge, it would be prudent to perform a weigh-in-motion study and thoroughly inspect the cover plate ends of the Warrior River Bridge for fatigue cracks.

IX. CONCLUSIONS

This chapter presented a probabilistic fracture mechanics method of calculating the risk of fatigue failure of a steel structure. With regard to highway bridges, it is very useful in determining how long one can extend the service life without exceeding a specified risk of failure. The need to extend the service life becomes increasingly important as bridges age and funds lack to replace them at the end of the 50-year design life. The method allows the engineer to compare how effective rerouting truck traffic, posting weight limits, and periodic inspections would be in limiting the risk of failure. Considering also the practicality and cost of each option, he can then determine the most viable option. While illustrated herein for the case of highway bridges, the method is general and can be applied to all types of steel structures.

ACKNOWLEDGMENTS

The work reported herein was sponsored by the Maryland Department of Transportation in conjunction with the Federal Highway Administration.

REFERENCES

1. **Albrecht, P., Wattar, F., and Sahli, A.,** Toward fatigue-proofing cover plate ends, Proc. W.H. Munse Symp. on Behavior of Metal Structures, ASCE National Convention, Philadelphia, Pa., May, 1983, 24.
2. **Albrecht, P. and Yamada, K.,** Rapid calculation of stress intensity factors, *J. Struct. Div. ASCE,* 103, No. ST2, 377, 1977.
3. **Ayyub, B. M. and Haldar, A.,** Practical structural reliability techniques, *J. Struct. Eng. ASCE,* 575, 1984.
4. **Fisher, J. W.,** Fatigue and Fracture in Steel Bridges — Case Studies, John Wiley & Sons, New York, 1984.
5. **Fisher, J. W., Albrecht, P., Yen, B. T., Klingerman, D. J., and McNamee, B. M.,** Fatigue strength of steel beams with welded stiffeners and attachments, NCHRP Report 147, Highway Research Board, National Research Council, Washington, D.C., 1972.
6. **Fisher, J. W., Frank, K. H., Hirt, M. A., and McNamee, B. M.,** Effects of weldments on the fatigue strength of steel beams, NCHRP Report 102, Highway Research Board, National Research Council, Washington, D.C., 1970.
7. **Fisher, J. W., Hausamann, H., Sullivan, M. D., and Pense, A. W.,** Detection and repair of fatigue damage in welded highway bridges, NCHRP Report 206, Highway Research Board, National Research Council, Washington, D.C., 1979.
8. **Harris, D. O., Lim, E. Y., and Dedhia, D. D.,** Probability of pipe fracture in the primary coolant loop of a PWR plant, U.S. Nuclear Regulatory Commission, Washington, D.C., 1981.
9. **Rippling, E. R. and Crosley, P. B.,** Narrative summary on fracture control in relation to bridge design, U.S. Department of Transportation, Federal Highway Administration, Washington, D.C., 1983.
10. **Rabbe, P. and Lieurade, H. P.,** Fracture mechanics study of crack growth rates in various steels, *Mem. Sci. Rev. Metall.,* 69, No. 9, 1972.
11. **Sahli, A. H. and Albrecht, P.,** Fatigue life of stiffeners with known initial cracks, 15th Natl. Symp. on Fracture Mechanics, ASTM STP 833, American Society for Testing and Materials, College Park, Md., 1984, 193.
12. **Shaaban, H. and Albrecht, P.,** Collection and analysis of stress range histograms recorded on highway bridges, Civil Engineering Report, University of Maryland, College Park, Md., 1983.
13. **Yazdani, N.,** Risk analysis of extending bridge service life, Ph.D. dissertation, Department of Civil Engineering, University of Maryland, College Park, Maryland, 1984.
14. **Yazdani, N. and Albrecht, P.,** Crack growth rates of structural steels in air and aqueous environments, Civil Engineering Report, University of Maryland, College Park, Md., 1983.
15. **Zettlemoyer, N.,** Stress concentration and fatigue of welded details, Ph.D. Dissertation, Lehigh University, Bethlehem, Pa., 1976.
16. **Albrecht, P. and Simon, S.,** Fatigue notch factors for structural details, *J. Struct. Div., ASCE,* 107, No. ST7, 1279, 1981.

Chapter 5

MODULARIZATION METHODS FOR EVALUATING FAULT TREES OF COMPLEX TECHNICAL SYSTEMS

Javier Yllera

TABLE OF CONTENTS

I. INTRODUCTION

Fault tree analysis is one of the most important techniques for system modeling and has been widely used in risk assessments together with event tree models. A fault tree is a graphic representation of the logical relationships between events in a system which result in a prespecified undesired state of it, defined as the top-event of the fault tree. The basic concepts for the deductive elaboration of a fault tree from a top-event are presented.[1]

A fault tree is a boolean reliability model, because every event can be described by a boolean variable, that takes the value 1 if the event occurs or 0 otherwise. Since fault tree gates work as boolean operators on their inputs, a fault tree describes a unique boolean function, called the structure function.

In order to quantify the fault tree, probabilities have to be assigned to its basic events (components). While small fault trees can sometimes be evaluated by hand, the evaluation of large fault trees requires the use of computerized methods. These can be of simulative (Monte Carlo) or analytical nature. Monte Carlo methods simulate the state of the fault tree components by means of random numbers adjusted to their life- and repair-time distributions, thus allowing one to know the state of the system at any time by solving the structure function.[2] Monte Carlo calculations do not permit the analysis of a fault tree independently from the failure data of its components; furthermore, they do not lead to any relationship between the reliability parameters of the system and those of its components either. In addition, results of Monte Carlo calculations have an stochastical character, and the number of trials needed to get accurate results increases with the inverse of the square of the value of the top-event probability. Therefore, the simulation of a highly reliable system requires too large a computing time. Reduction of this computing time can be achieved by applying weighting techniques.[3] However, Monte Carlo simulation can handle complex boundary conditions, e.g., restricted repair capacity or cold reserve systems, with ease, whereas analytical methods can only handle them in a simplified way or not at all.

The most frequently used analytical methods for fault tree analysis are based on the determination of the minimal cut sets (MCSs). These are the smallest conjunctions of basic events that lead to the top-event, and are independent from the failure data. The examination of the MCSs provides in formation about failure modes and weak points of the system (MCSs with few components, components appearing in many MCSs, etc.) and can possibly suggest ways by which the system could be improved. The calculation of reliability parameters for the system given those of its components is straightforward once the MCSs are known, but the determination of the MCSs can be a difficult task. Some efficient algorithms have been developed for this purpose.[4,5] However, the determination of all MCSs of a tree is many times impossible in practice, because the corresponding computational effort cannot be limited by a polynomial of the number of tree events.[6] These kind of computational problems are called "NP-hard" (nonpolynomial hard). Therefore, the determination has to be limited to those MCSs that contribute substantially to the top-event, i.e., a cutoff procedure has to be used. On the other hand it would be impossible to keep track of some thousands or more MCSs.

A cutoff procedure can be applied depending either on the cut set length (qualitative cutoff) or on some cut set failure parameters, e.g., its unavailability (quantitative cutoff). A qualitative cutoff procedure may lead to considerable errors, especially if the components have quite different failure data. Another method of selecting cut sets in some order of probabilistic importance is to simulate the system by the Monte Carlo method and to obtain one or more MCSs each time the system fails. Since a reliable system does not fail during most of the trials, weighting methods or prolongation of the observation period under neglection of maintenance have to be used in order to reduce the computing time. The major disadvantage of this method is that the contribution of the undiscovered MCSs cannot be estimated. A better approach consists in a modularization of the fault tree that permits a separate evaluation of those tree branches without common events in the remainder of the tree. This leads to a drastic reduction of the MCSs to be investigated. Unfortunately, the modularization is often strongly restricted by the presence of shared system units, e.g., support systems, that result in interlinked fault tree branches. This article proposes a more general module concept which allows for an efficient modularization, even in such cases, and gives a method for selecting and evaluating them. The evaluation algorithm presented here is only valid for coherent systems (i.e., systems that do not deteriorate when the reliability characteristics of their components improve) although it would not be too difficult to adapt it to noncoherent systems.

II. MODULAR DECOMPOSITION OF FAULT TREES

Modules have been long considered in many theoretical and technological applications as assemblies of components within a system that can be treated as an independent subsystem and thus can be, for instance, removed and replaced as a whole. A formal module definition of a coherent binary system was first given by Birnbaum and Esary as follows.[7] Let (C,φ) denote a binary system given by a coherent structure function φ operating on the component set $C = \{c_1, c_2, \ldots, c_n\}$, whose state is indicated by the boolean vector $\underline{x} = (x_1, x_2, \ldots x_n)$. A nonempty subset A of C is a modular set if $\varphi(\underline{x})$ can be expressed as:

$$\varphi(\underline{x}) = \psi(\alpha(\underline{x}^A), \underline{x}^{C-A}) \tag{1}$$

where α is a coherent structure function on A, called the modular function, and ψ is another coherent structure function on α and the remainder set of A. When A is a modular set, the subsystem (a,α) is said to be a module of the system (C,φ).

A trivial conclusion to be drawn from the above definition is that every component as well as the system as a whole may be considered as a module. For differentiation all but the aformentioned modules are called proper modules. A coherent boolean function that does not have proper modules is called a "prime function". For instance, the structure function of an N out of M system is prime.

A modular decomposition of a coherent binary system (C,φ) is defined as a partition of the system in a disjoint set of modules, $(A_1, \alpha_1), (A_2, \alpha_2), \ldots, (A_n, \alpha_n)$:

$$\varphi(\underline{x}) = \psi(\alpha_1(\underline{x}^{A_1}), \alpha_2(\underline{x}^{A_2}), \ldots, \alpha_n(\underline{x}^{A_n})) \tag{2}$$

A module is another coherent binary system, which might be decomposed itself in smaller modules, thus leading to a hierarchical decomposition of the system in modules, submodules, etc. In a reliability block diagram, a module can be represented as a two-terminal subblock, i.e., a subdiagram that has only one input from the remainder of the diagram and only one output to it. In a fault tree representation, every branch whose events do not appear outside of it is a module. For example, in the fault tree of Figure 1 the components belonging to the tree branch under gate G_1 are a modular set, as the structure function describing the top-event can be rewritten as $TOP = G_1 \wedge x_{12}$, G_1 representing the structure function for the subtree under this gate, i.e., the modular function.

In a fault tree without repeated events, every branch represents a module. When a fault tree contains repeated events an exhaustive modularization of it may become a difficult task. For coherent systems, the relationship between MCSs and modules is completely clear.[7] Therefore, a criterion can be established for deciding whether a set of components is a

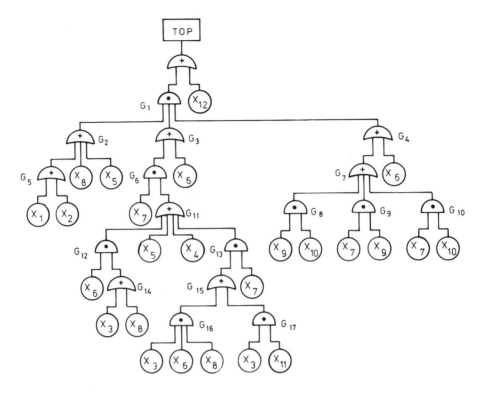

FIGURE 1.   Fault tree example 1.

modular set or not. This relationship is, however, not useful for the purpose referred to in this chapter, because the MCSs of the fault tree are not available. In addition, it should be decided first, which component sets have to be checked for modularity.

According to Equation 2, several modular decompositions of a coherent system might be obtained. However, a very important result of this module definition is that the modular factorization of a coherent binary system produces a unique decomposition into, in a qualified sense, its largest possible disjoint modules, also called modular factors. The function which organizes these modules is either a pure disjunction, a pure conjunction, or a prime boolean function. This statement is equivalent to the unique factorization theorem for simple games, proved by Shapley.[8] This should not be surprising if one takes into account that a simple game can be clearly and unequivocally described by a unique set of minimal winning coalitions; and in the same way, a coherent boolean function has a unique representation in terms of its MCSs. Both minimal winning coalitions and MCSs fit in the global definition of a clutter. A clutter can be defined as a family of different subsets of a finite set, having the property that no subset of the family is properly contained in another one. Obviously, the MCSs of a structure function are a clutter on the set of its components. An algorithm for the decomposition of clutters have been developed by Billera.[9] Such an algorithm can also be applied to the decomposition of a coherent boolean function, once its minimal cut sets and path sets are known.[10] The algorithm has been reformulated and improved by Chatterjee for its application to the modular decomposition of coherent fault trees.[11]

If the above defined modular factors, which are also coherent binary systems, are decomposed recursively into their own modular factors until further decomposition becomes impossible (i.e., the modular factors are single components), then the so-called finest modular decomposition has been reached. Chatterjee's algorithm leads to this representation of the fault tree, which can also be defined in a more usual fault tree terminology as an equivalent

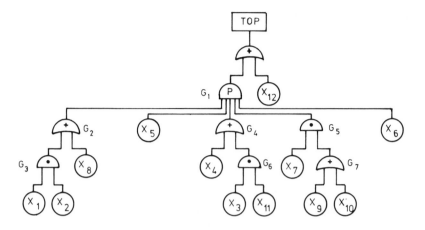

$$G_1 = (X_5 \lor G_2) \land (X_6 \lor G_5 \land (G_4 \lor X_5))$$

FIGURE 2.   Finest modular decomposition of the fault tree example given in Figure 1.

representation of the tree, where (1) every branch of the tree is independent, i.e., there are no repeated events, and therefore every subtree can be considered as a module and (2) the logical function associated with each gate is either an OR (disjunction) or an AND (conjunction), having no inputs from other gates of the same type, or a prime boolean function.

The finest modular representation of the fault tree of Figure 1 is given in Figure 2. Since Chatterjee's algorithm produces this representation starting from the MCSs, the prime gates are characterized by a sum of MCSs based on their inputs, but in general, a prime gate would not be different from a boolean function representing a fault tree without modules. The basic events of this tree (some of them are necessarily repeated) are the inputs of the original prime gate. This algorithm is not useful for the purpose of reducing the effort in computing system reliability parameters, because it requires all MCSs as input information. Once these are available, the desired parameters can be obtained directly and no other representation is needed. In many cases, however, it is impossible in practice to obtain all MCSs.

If we define by $N(\varphi)$ a measure of the computational effort required for the evaluation of a function φ, it is easy to show by considering the intuitive module concept of an independent subsystem which can be evaluated separately and treated like a single component, that for a given modular decomposition:

$$N(\varphi) \geqq N(\psi) + \sum_{i=1}^{n} N(\alpha_i) \tag{3}$$

because the effort for evaluating a function grows nonlinearly with the number of components. From this point of view, the finest modular representation would be the ideal representation of a fault tree, which in fact, it is. On the other hand, the effort involved in the modularization process has to be taken into account too. Until now, no suitable method has been proposed for obtaining the finest modular representation of a fault tree directly from its structure. Nevertheless, there exist some algorithms which perform an effective modularization by recognizing independent subtrees and groups of unrepeated events which are inputs to the same gate, and by changing the tree structure without altering its truth table in order to produce such patterns with an admissible effort.[12,13] However, they do not normally lead to the finest modular representation. In the following, one of these algorithms is presented.

A. Algorithm for the Modular Decomposition of Coherent Fault Trees

The algorithm for modularization consists of two parts. First, the system structure is handled and optimized. During this process, independent subtrees are recognized. Second, the modularization is carried out.

1. Fault Tree Structure Manipulations

Manipulations of the fault tree structure are performed in the following steps.

1. Gates are numbered in such an order that the number of a gate is always greater than the number of its inputs, allowing one to investigate some gate properties based on those of their inputs just by visiting the gates in the order 1,2, . . . ,n. In addition, events not covered by the top are eliminated and gates with a single input are bypassed.
2. A threshold value ϵ_a is given for some performance measure (unavailability or failure frequency density), and components that are below this value are eliminated as well as the intermediate gate events that become impossible, given that these components never fail. The contribution of these components to the truncation error is computed as if each of them was a MCS. If, for example, component x_6 of Figure 1 never fails, then itself and also gates G_{12} and G_{16} are deleted everywhere in the tree.
3. The number of repetitions of each event is computed, and independent subtrees, i.e., subtrees without common events with the rest of the tree, are determined and marked as module candidates. Unrepeated gates being inputs to gates of the same type (only AND, OR) are coalesced except for those marked earlier as module candidates.
4. Each gate is now visited in the order 1,2, . . . ,n and some possible simplifications are investigated. These are performed if the following patterns occur:

$$G = (E_i \lor E_j) \land (E_i \lor E_k) \land \ldots \text{ is converted into}$$

$$G = E_i \lor (E_j \land E_k \land \ldots)$$

and

$$G = (E_i \land E_j) \lor (E_i \land E_k) \lor \ldots \text{ is converted into}$$

$$G = E_i \land (E_j \lor E_k \lor \ldots)$$

i.e., the event E_i is shifted to higher levels. For example, component x_3 can be eliminated from gates G_{16} and G_{17}, and inserted as input of gate G_{13}.

$$G = E_i \land \beta(E_i,\ldots) \land \ldots$$

where $\beta(E_i, \ldots)$ represents the subtree under an unrepeated gate containing event E_i, is converted into

$$G = E_i \land \beta(E_i = 1,\ldots) \land \ldots$$

and

$$G = E_i \lor \beta(E_i,\ldots) \lor \ldots \text{ is converted into}$$

$$G = E_i \vee \beta(E_i = 0,...) \vee \ ...$$

(State 1 means failed and 0 intact.)

For example, it is possible to eliminate component x_7 as input of gate G_{13}.

These two transformations are achieved using some routines of the step 2. Care is always taken not to change gates appearing elsewhere in the tree.

The gates that have been modified are tested for independence from the rest of the tree, and if so, marked as module candidates. The resulting fault tree serves as input to the modularization process.

2. Modularization

Modularization is performed in a bottom-up tree traversal, and the modules are numbered in the order in which they are created. Each gate is visited and its inputs investigated. Unrepeated components or module inputs are collected in a module, whose modular function is given by the logic of the gate. This kind of module is called an "easy module" because its evaluation is straightforward. If all inputs are modularized in this way, the created module substitutes the gate everywhere in the tree. If no module is found or any inputs are left, and if the gate has been marked before as module candidate, then the subtree under the gate is converted into a module. This kind of module is called a "gate module" and it has to be evaluated in terms of MCSs.

B. Modular Decomposition of Noncoherent or Multistate Fault Trees

For noncoherent systems, Equations 1 and 2 can be used with the introduction of slight modifications, namely with the condition that both intact and failed states of a component belong to the same modular set. Obviously, either some modular function or the organizing function, or both, will be noncoherent. It is well known, that for coherent systems, better probability upper bounds are available than for noncoherent ones. Therefore, it is convenient to reduce the number and size of noncoherent functions as much as possible. (For a more detailed discussion of this see Chu and Apostolakis.[14]) For a noncoherent system, no factorization theorem has been given and Chatterjee's algorithm is not applicable, because a noncoherent system can be described by more than one family of prime implicants.

The same definition can also be extended to systems with multistate components, always under the condition that a modular set has to include every possible state of its components. The modular and organizing functions should be checked for coherence in order to establish which type of probabilistic upper bound is applicable to each function. A method to recognize whether a system with multistate components is coherent or not has been given by Caldarola.[15]

III. A MORE GENERAL DECOMPOSITION METHOD

Now we give a much broader module definition, which also covers the module definition according to Equation 1. Let (C,φ) be a structure function. Each nonempty subset A of C which permits the structure function to be expressed as

$$\varphi(\underline{x}) = \psi(\alpha_1(\underline{x}^A), \alpha_2(\underline{x}^A),...,\alpha_n(\underline{x}^A), \underline{x}^{C-A} \tag{4}$$

n being a finite number and $\alpha_1, \alpha_2, \ldots , \alpha_n$ are functions dependent only on the subset A, is a modular set of (C,φ). A module is defined by a modular set and a finite number of modular functions, $(A, \alpha_1, \alpha_2, \ldots , \alpha_n)$.[16] If $n = 1$, Equation 4 is immediately reduced to Equation 1. From now on, the name "module" will be reserved for first order modules,

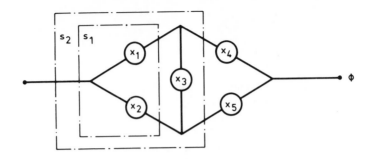

FIGURE 3.   Reliability block diagram of the bridge structure.

i.e., n = 1, and any other modules of order n > 1 will be designated as "supermodules". As before, ψ is the organizing function. It is obvious, that every set of components can be a modular set by defining an adequate set of modular functions.

In a similar way to Equation 2, a modular decomposition can be defined as

$$\varphi(\underline{x}) = \psi(\alpha_1^1(\underline{x}^{A_1}), \alpha_2^1(\underline{x}^{A_1}),\ldots,\alpha_{n_1}^1 \underline{x}^{A_1}),$$

$$\alpha_1^2(\underline{x}^{A_2}), \alpha_2^2(\underline{x}^{A_2}),\ldots, \alpha_{n_2}^2(\underline{x}^{A_2}),$$

$$\alpha_1^m(\underline{x}^{A_m}) \alpha_2^m(\underline{x}^{A_m}),\ldots,\alpha_{n_m}^m(\underline{x}^{A_m})) \qquad (5)$$

where, A_1, A_2, . . . , A_n are disjoint subsets of C. Modular functions from different supermodules are statistically independent. However, for a given supermodule $(A_i,\alpha_1^i,\alpha_2^i, \ldots, \alpha_{n_i}^i)$ the functions $\alpha_1^i,\alpha_2^i, \ldots, \alpha_{n_i}^i$ are normally not statistically independent. This fact has to be considered when quantifying the organizing function ψ.

In a reliability block diagram, a module was a subblock having only one input from the remainder of the diagram and one output to this remainder. A supermodule has only one input from the remainder of the system and a finite number of outputs to it. In order to illustrate this, consider, e.g., the bridge structure, Figure 3, which is known to be prime (i.e., it has no conventional proper modules according to Equation 1). The block S_1 containing components x_1 and x_2 could be replaced by a supermodule. This would not be meaningful, however, because, if we attempt to describe its outputs by modular functions, we would just rename the basic events x_1 and x_2. On the other hand, if the block S_2 containing components x_1, x_2, x_3 is considered as a supermodule, the reliability block diagrams for the organizing and modular functions change as shown in Figure 4.

In a fault tree, representation of the structure function a supermodule is just a part of the tree containing all the components belonging to the modular set and having several outputs to the rest of the tree. The fault tree representation of a bridge structure with the selected supermodule $(S_2, \alpha_1, \alpha_2)$. $\underline{x}^{S_2} = (x_1, x_2, x_3)$ constructed in accordance with the functions shown in Figure 4 is given in Figure 5.

A. Some Properties of Supermodules

The application of pivotal decomposition on all the modular functions of a supermodule leads to a sum of incompatible products, each of them decomposable in two statistically independent factors. When $(A,\alpha_1,\alpha_2, \ldots,\alpha_n)$ is a supermodule of (C,φ) the application of pivotal decomposition on α_1, α_2, . . . , α_n to the organizing function ψ leads to:

$$\varphi(\underline{x}) = \psi(\alpha_1 = \alpha_2 =\ldots= 0, \underline{x}^{C-A}) \wedge \overline{\alpha}_1 \wedge \overline{\alpha}_2 \wedge\ldots\wedge \overline{\alpha}_n \vee$$

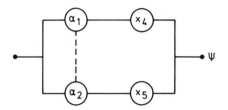

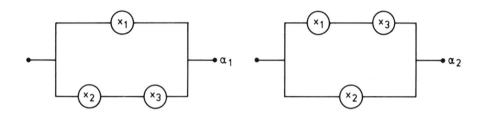

FIGURE 4.   Reliability block diagrams of the functions α_1, α_2, and ψ.

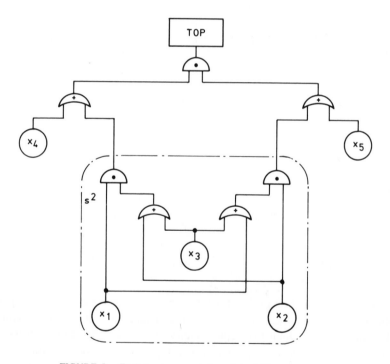

FIGURE 5.   Fault tree representation of the bridge structure.

$$\bigvee_{i=1}^{n} \left\{ \psi(\alpha_i = 1, \alpha_s = 0, \underset{\substack{s=A_j,\ldots,n \\ s\neq i}}{\underset{\longrightarrow}{x}}^{C-A}) \wedge \alpha_i \bigwedge_{\substack{s=\wedge \\ s\neq i}}^{n} (\wedge \, \overline{\alpha}_s) \right\} \vee$$

$$\bigvee_{1\leq i<j\leq n}^{\binom{n}{2}} \left\{ \psi(\alpha_i = \alpha_j = 1, \alpha_s \underset{\substack{s=A_1\ldots,n \\ s\neq i,j}}{=\ldots=} 0, \underset{\longrightarrow}{x}^{C-A}) \wedge \alpha_i \wedge \alpha_j \wedge \left(\bigwedge_{\substack{s=\wedge \\ s\neq i\neq j}}^{n} \overline{\alpha}_s \right) \right\} \vee$$

$$\vee \, \psi(\alpha_1 = \alpha_2 = ,\ldots, \alpha_n = 0, \underset{\longrightarrow}{x}^{C-A}) \wedge \alpha_1 \wedge \alpha_2 \wedge\ldots\wedge \alpha_n \qquad (6)$$

In this expression $\alpha_i(\underline{x})^a$ was replaced by α_i for simplicity. $\overline{\alpha}_i$ means the complement or negation of α_i. Equation 6 can be easily proved and underlines the idea that the components of a supermodule can only influence the performance of the system through the performance of the modular functions. In other words, the state of the system depends only on the states of the modular functions and not on the combination of the states of the components of the modular set which led to them. In a module, the unique modular function and the organizing function are uniquely determined, once the modular set is defined. In supermodules, several sets of modular functions can be defined on the same modular set of components. For instance, in the case of the bridge structure, presented in Figure 3, the following modular functions

$$\alpha_1 = x_1 \wedge (x_2 \vee x_3), \ \alpha_2 = x_1 \vee x_3, \ \alpha_3 = x_2$$

can also be defined for the modular set S_2, $\underline{x}^{S2} = (x_1, x_2, x_3)$ together with the organizing function

$$\psi = (x_4 \vee \alpha_1) \wedge (x_5 \vee \alpha_2 \wedge \alpha_3)$$

Every combination of n components can be considered as a modular set, because there always exists the possibility of defining n modular functions each of them depending on only one component of the modular set, e.g., the supermodule S_1 of Figure 3, but this kind of supermodule is trivial and does not offer any advantage. A supermodule is said to be nontrivial if the number of modular functions is smaller than the number of components included in the modular set.

Finally, if $(A, \alpha_1, \alpha_2, \ldots, \alpha_n)$ is a supermodule of (C, φ), then $(A, \alpha_1^D, \alpha_2^D, \ldots, \alpha_n^D)$ is also a supermodule of (C, φ^D), where $\alpha_1^D, \alpha_2^D, \ldots, \alpha_n^D, \varphi^D$ are the dual functions of α_1, $\alpha_2, \ldots, \alpha_n, \varphi$, respectively.

B. Evaluation of Supermodules

While the evaluation of fault trees produced by decomposition in normal modules can be carried out by normal cut set enumeration methods, an evaluation method for fault trees decomposed in supermodules has to take into account that modular functions of the same supermodule are normally statistically dependent. Therefore, conjuctions of such modular functions that appear at some cut set representation cannot be evaluated as the product of expected values of the single modular functions, i.e.,

$$E\{\alpha_i^1 \wedge \alpha_j^1\} \neq E\{\alpha_i^1\} \cdot E\{\alpha_j^1\}$$

This case will appear in the evaluation of the organizing function or in the evaluation of the modular functions if the corresponding supermodule has been decomposed in further supermodules. Equation 6 can be used as a possibility for eliminating these statistical dependencies from the organizing function. A more efficient method developed by the author will be shown by handling an example.[16]

Let us consider the fault tree in Figure 6 for explanation purposes. This fault tree could also represent a module found by some algorithm like the one presented above. In this tree, the supermodule of 2nd order (A, α_1, α_2), $\alpha_1 \equiv G_5$, $\alpha_2 \equiv G_7$, has been selected for simplicity but without loss of generality. The determination of the MCSs for the corresponding organizing function would lead to the following representation of the structure function:

$$\varphi(\underline{x}) = x_1 \wedge x_2 \wedge x_6 \vee x_1 \wedge x_2 \wedge x_7 \wedge \alpha_2 \vee$$

$$x_1 \wedge x_6 \wedge \alpha_1 \vee x_1 \wedge x_7 \wedge \alpha_1 \wedge \alpha_2 \tag{7}$$

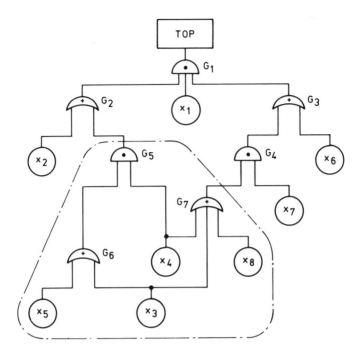

FIGURE 6.   Fault tree example 2.

From this representation the different modular functions or products of them can be determined and quantified separately in terms of their MCS representations:

$$\alpha_1 = x_4 \wedge x_5 \vee x_3 \wedge x_4$$

$$\alpha_2 = x_3 \vee x_4 \vee x_8$$

$$\alpha_1 \wedge \alpha_2 = x_4 \wedge x_5 \vee x_3 \wedge x_4 \tag{8}$$

Insertion of the results obtained for the functions G_5, G_7, and $G_5 \wedge G_7$ in the organizing function, Equation 7, allows an estimate to be obtained of the desired system value. However, it is possible by this procedure, that some nonminimal or repeated cut sets are considered in the calculation. For instance, the organizing function may contain cut sets like $x_i \wedge x_j \wedge \alpha_r$, $x_i \wedge x_j \wedge \alpha_s$, which only differ by the modular functions α_r, α_s and in addition, α_r and α_s have some common MCS. If upper bounds corresponding to the quality standards of MCS enumeration are desired, a demodularization of the supermodules can be performed and the nonminimal cut sets can be eliminated. The demodularization is the reverse problem of the modularization. By this process modular functions or products of them contained in the MCSs of the organizing function are replaced by their corresponding MCS representation. As a result, the MCSs of the original structure function are obtained. This process can be carried out with low computing time requirements, because modular functions or products of them have been already quantified and a cutoff procedure is very efficient in this case. Their replacement by their MCSs takes place, in addition, in a single development state.

The major advantage of this method is that the computational effort is never greater than the effort for a normal cut set enumeration for every possible selection of supermodules. This means, that there is no practical limitation to the order of the supermodules to be selected, since only the products of modular functions appearing in the MCSs of the organizing function are evaluated. Application of Equation 6 requires the evaluation of 2^n

products of modular functions instead. In addition, these products represent noncoherent functions, except when the original tree is coherent. A more detailed description of the evaluation method presented and some others are given in Reference 16.

C. Selection of Supermodules

Now, we consider the problem of selecting those supermodules which appreciably reduce the effort required for evaluating fault trees by the method presented in the last section. It is obvious that the effort for finding the supermodules has to be taken into account in order to establish a valid criterion. In considering this question, the possibility of using cutoff procedures should be ignored, because the failure data influence could change every prediction based on structural considerations of the fault tree. The potential number of cut sets of a structure function and their respective length will be considered as an estimate of the computational effort involved in its probabilistic evaluation.

Decomposition in supermodules is particularly interesting in the case of prime functions and functions where no standard modules can be found with the searching methods available. These kinds of functions have to be evaluated hitherto by MCS enumeration methods. If a reduction in the computational effort for their evaluation can be achieved by using super modules, then the advantages of their use have been proved for any kind of functions, because standard modular decomposition can be used too, and this process leads to one of the types of function described above.

An overproportional increase of the number of MCSs of a tree is originated by "AND" gates having inputs with more than one potential cut set. In this case, the problem can be reduced to linearizing the effort in evaluating the AND gates. Let us consider an AND gate with inputs G_1, G_2. This is not a real restriction on the number of inputs, since G_1 or G_2 could in turn represent AND gates. If the output event of gate G_1 is considered as a modular function α_1 of a supermodule, whose modular set consists of the components contained in the tree branch under gate G_1, then the remaining modular functions can be defined as those events in the tree branch under G_1 that are inputs to the other gates outside of the branch. If such events do no exist, a normal module has been found. In a general case, there would be some modular functions $\alpha_r, \alpha_s, \ldots$, which do not enter gates in the subtree under G_2, and some other modular functions $\alpha_i, \alpha_j, \ldots$, which at least enter gates in the subtree under G_2. At the moment, our attention is paid to the latter. This situation is indicated in Figure 7.

In order to establish a relationship between the computational effort involved in the evaluation of gate G with and without modularization, a top-down development of gate G is performed, stopping when those events are found which have been defined as modular functions. The resulting sum of products can be expressed as:

$$G \equiv \kappa_0 \wedge \alpha_1 \vee \kappa_i \wedge \alpha_1 \wedge \alpha_i \vee \kappa_j \wedge \alpha_1 \wedge \alpha_j \vee \ldots \vee \kappa_{ij} \wedge \alpha_1 \wedge \alpha_i \wedge \alpha_j \vee \ldots \quad (9)$$

for $i \neq j \neq 1$ and $\alpha_1 \equiv G_1$.

In this expression, the functions κ represent families of cut sets formed with components which do not belong to the modular set. The computational effort, measured for instance as the potential number of cut sets, for the evaluation of the subtree of gate G by a cut set enumeration algorithm can be estimated as:

$$N(G) = N(\kappa_0) \cdot N(\alpha_1) + N(\kappa_i) \cdot N(\alpha_1) \cdot N(\alpha_i) +$$

$$N(\kappa_j) \cdot N(\alpha_1) \cdot N(\alpha_j) + \ldots +$$

$$N(\kappa_{ij}) \cdot N(\alpha_1) \cdot N(\alpha_i) \cdot N(\alpha_j) + \ldots \quad (10)$$

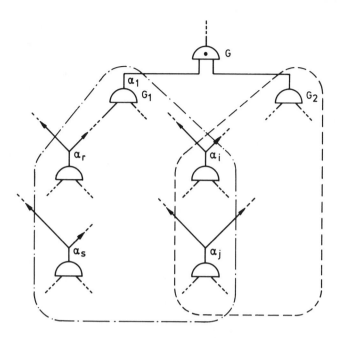

FIGURE 7.   Illustrating scheme for the supermodule selection strategy.

If, on the other hand, the fact that the functions κ are statistically independent of modular functions is taken into account, the evaluation of a product like $\kappa_i \wedge \alpha_1 \wedge \alpha_i$ can be split into the separate evaluation of κ_i and $\alpha_1 \wedge \alpha_i$. Hence, the computational effort $N^*(G)$ for the probabilistic evaluation of gate G considering the supermodule would be:

$$N^*(G) = N(\kappa_0) + N(\alpha_1) + N(\kappa_i) + N(\alpha_1) \cdot N(\alpha_i) +$$

$$N(\kappa_j) + N(\alpha_1) \cdot N(\alpha_j) + ... +$$

$$N(\kappa_{ij}) + N(\alpha_1) + N(\alpha_i) \cdot N(\alpha_j) + ... \tag{11}$$

where it has been assumed that the function α_1 has to be evaluated only once. A comparison of equations 10 and 11 reveals that the computational effort using modularization decreases substantially as the number of potential cut sets of κ functions departs from one. This reasoning can be applied to conjunctions between the other modular functions α_r, α_s, ... and other gate events outside the tree branch under gate G as well. Such conjunctions can exist explicitly as an AND gate in the given form of the tree or can represent an intermediate stage in a cut set enumeration algorithm.

According to the above arguments, the selection of supermodules can be carried out by investigating the "AND" and "N out of M" gates with more than one gate input following a decreasing order to their potential number of cut sets. A supermodule is constructed with the subtree of the input which has the largest possible number of cut sets, unless the modular set is not disjoint from the modular sets of supermodules selected before making this impossible. The remaining modular functions are defined again as those events of the subtree which are the input of gates outside the subtree.

When no supermodules can be found this way, the procedure is applied within each supermodule. It then leads to a set of supermodules of a lower hierarchical level and is continued until no more nontrivial supermodules can be found.

This selection procedure is likely not to be an optimal strategy in many cases, but its

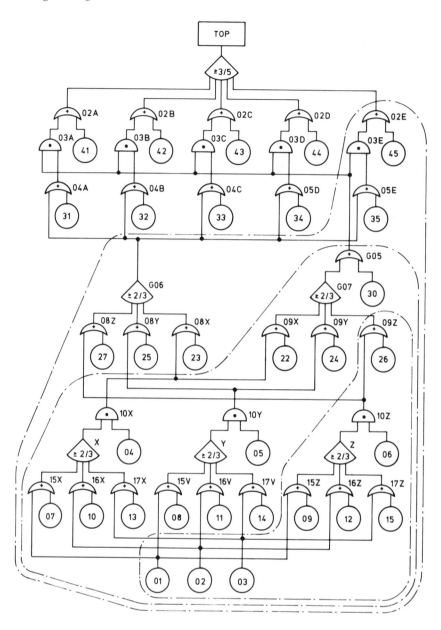

FIGURE 8.   Fault tree example 3.

efficiency has been justified by the above reasoning. It can be accomplished in a very fast and easy way, whereas the application of some other procedures might involve time-consuming searching procedures. The selection procedure proposed will be illustrated now using the fault tree of Figure 8. This fault tree was presented as a benchmark example of tree without normal modules in Reference 17. It will be evaluated for demonstration purposes in the section V as well. The first supermodule selected according to the strategy proposed, consists of the elements of the tree branch under gate 02E, which is one of the gates with the maximum number of potential cut sets entering an AND gate (an "N out of M" gate is a disjunction of "AND" gates). The modular functions are characterized by the outputs of gates 02E, G06, and G05 (a third order supermodule). No other disjoint supermodules can be found by the algorithm at this hierarchical level. The application of the selection algorithm within the supermodule thus obtained gives the supermodule under gate G05 with modular functions equivalent to the outputs of gates G05, 10X, 10Y, and 10Z. No more disjoint supermodules are found at the second hierarchical level. At the third hierarchical level the supermodule under gate 09Z with modular functions 09Z, 10Z, 01, 02, and 03 is selected. No other disjoint supermodules can be found by the algorithm. The algorithm stops at this hierarchical level, because modularization of the subtree under gate Z would not reduce the evaluation effort, since no other gates enter gate 10Z, and the supermodules under gates 15Z, 16Z, and 17Z are trivial.

IV. ALGORITHM FOR THE MODULAR EVALUATION OF FAULT TREES

The algorithm which is presented now has been implemented in the computer code TREEMOD,[13] which forms part of the program system RISA.[18]

After a normal modular decomposition, a hierarchical set of independent subtrees is produced. The modules can be evaluated in the same order in which they were created by the algorithm presented in Section II.A. In case of a fault tree corresponding to an "easy module", the corresponding structure function is a pure conjunction or a pure disjunction. Its unavailability $U_m(t)$ and failure frequency density $h_m(t)$ can be calculated by the following equations:

OR module

$$h_m(t) = \sum_{j=1}^{n} h_j(t)$$

$$U_m(t) = 1 - \prod_{j=1}^{n} (1 - U_j(t)) \tag{12}$$

AND module

$$h_m(t) = \sum_{j=1}^{n} h_j(t) \cdot \prod_{\substack{k=1 \\ j \neq k}}^{n} U_k(t)$$

$$U_m(t) = \prod_{j=1}^{n} U_j(t) \tag{13}$$

Otherwise the corresponding independent subtree, e.g., a "gate module", which should be evaluated usually by a direct determination of its MCSs, is decomposed in supermodules according to the selection procedure explained in section III.C. The independent subtree is then evaluated, using the method presented in section III.B. The evaluation algorithm, which uses a top-down development procedure for the determination of MCSs of a given function, follows the depth first sequence.[19] That means, the algorithm works recursively. It can be summarized in the following way:

1. Select supermodules.
2. Call "development procedure (top event,0)".
3. Perform demodularization an minimalization in order to get the MCSs of the independent subtree.
4. End.

Development procedure (i,n): Estimates reliability parameters for event or product of events (i).

1. Check if the procedure was called before with the same parameters and, if true, return former results.
2. Set stopping conditions defined as the modular functions of the supermodules of the hierarchical level $n + 1$.
3. Search for previous results of the procedure for the largest possible subset j of i at the hierarchical level n.
4. Perform a top-down development of set i until stopping conditions are found. If a subset j was found at step (3), then use the results for j in speeding up the development of set i. A family K of partially and completely developed cut sets will be obtained.
5. For all supermodules at the hierarchical level $n + 1$, search and sort the different modular functions or products of modular functions of the same supermodule that are present in the cut sets of family K, and form with them the family J.
6. For all $j \in J$ call "development procedure $(j, n + 1)$" and use the results obtained in order to recalculate $k \in K \mid j \subseteq k$.
7. Obtain reliability parameters for the set i from the reliability parameters of the members of family K.
8. Store family K and reliability parameters of set i for future use in the demodularization process and in steps (1) and (2).
9. End.

The most complicated and also the most important part of this algorithm is that of step (5). The efficiency of the top-down development performed in it has a big influence on the total computing time. Therefore, it is important to optimize all the variables affecting the efficiency of this part of the algorithm. Notice, that in the case that "i" is the top event of a tree and no stopping conditions are given, the algorithm for this step can be used as a normal top-down algorithm for enumerating cut sets. The top-down algorithm used in code TREEMOD for this purpose is presented in the next section.

A. Top-down Algorithm

Top-down algorithms generate cut sets by substituting each gate by its inputs, beginning with the top gate. While substitution of AND gates increases the length of the cut sets, substitution of an OR gate produces new cut sets differing only in one of their events. Thus, a matrix is created in such a way that each of its rows is a cut set after all gates have been substituted by components. If the tree contains replicated events, then the cut sets obtained have to be checked for minimality. In a more general version of this algorithm the starting point can be a given set of events rather than the top event, and substitutions are only performed until stopping conditions are found. The efficiency of the algorithm depends on the total number of substitutions performed. Despite the drastic reduction in the number of cut sets achieved by the modularization methods explained, the top-down development can lead to a very large number of cut sets in many cases. Therefore, a probabilistic cutoff procedure has to be used for such trees, since limitations in storage capacity and running time do not allow one to handle them otherwise. On the other hand, normally, a reduced number of cut sets has a significant influence on the system reliability parameters. For this reason an efficient cut off procedure is an important tool for reducing the number of substitution steps and saving a great amount of computational work. In code TREEMOD, two cutoff levels, an absolute ϵ_a and a relative ϵ_r, are used to eliminate as early as possible those cut sets which have a preselected reliability parameter (failure frequency density or unavailability), either less than ϵ_a or less than the product of ϵ_r and the contribution of the cut sets already obtained at that time.

A conservative estimate of the unavailability $U_i(t)$ or failure frequency density $h_i(t)$ of a partially developed set i can be computed from the reliability parameters of the components it contains. Standard modules are considered as components for this purpose. Should the set also contain modular functions of supermodules or products of them that have been quantified already, they are taken into account for the estimate too. On the other hand, probabilities of one are assumed for any other parts of the set, i.e., event gates or non-quantified modular functions or products of them. The failure frequency density and unavailability of a set i estimated according to their quantified parts 1,2, . . . ,j are calculated as follows:

$$U_i^j(t) \; = \; \prod_{k=1}^{j} U_k(t) \tag{14}$$

$$h_i^j(t) \; = \; \prod_{k=1}^{j} h_k(t) \cdot \prod_{\substack{l=1 \\ l \neq k}}^{j} U_l(t) \tag{15}$$

If the estimate of a set is smaller than one of the cutoff criterions, then the set is eliminated and this estimate is used to calculate a conservative estimate of the total contribution of all eliminated sets to the desired system parameter.

The cutoff procedure can be made efficient by reaching the level of components or stopping conditions in few substitutions steps, since a cut set cannot be truncated as long as it only contains gates. Given the numeration sequences of gates explained in section II.A.1, the probability that a gate has component inputs grows as its number decreases.

Consider an insignificant set with components x_1 and x_2 and gates G_i and G_j that do not yet allow a truncation. For this explanation, it is assumed that G_i has only component inputs and G_j only gate inputs. The substitution of one of these gates will produce the following sets depending on the gate logic.

For "AND" gates

$$x_1 \wedge x_2 \wedge x_{i_1} \wedge x_{i_2} \wedge ... \wedge x_{i_n} \wedge G_j$$

or

$$x_1 \wedge x_2 \wedge G_i \wedge G_{j_1} \wedge G_{j_2} \wedge ... \wedge G_{j_m}$$

and for "OR" gates

$$x_1 \wedge x_2 \wedge x_{i_1} \wedge G_j$$
$$x_1 \wedge x_2 \wedge x_{i_2} \wedge G_j$$

$$x_1 \wedge x_2 \wedge x_{i_n} \wedge G_j$$

or

$$x_1 \wedge x_2 \wedge G_i \wedge G_{j_1}$$
$$x_1 \wedge x_2 \wedge G_i \wedge G_{j_2}$$

$$x_1 \wedge x_2 \wedge G_i \wedge G_{j_m}$$

Substitution of gate G_j does not allow one to neglect any set in both cases, as the estimates according to Equation 14 and 15, respectively, do not change, whereas if G_i is substituted truncation may be possible. If the substitution sequence is G_j, G_i, then more time is spent eliminating the sets and if G_j is an OR gate, then the estimate of the cut sets truncated is m times greater than in the case of sequence G_i, G_j, m being the number of inputs of gate G_j.

A development sequence that always substitutes the gate with the lowest number existing in the set can be followed by using a computational stack. The gates are stored in the stack beginning with the highest number. The gate at the top of the stack always has the lowest level and is developed first. None of its inputs can be already in the stack, because of the special numeration used. If new gates have to be added to the sets, they would be stored at the top of the stack. When the stack is empty, the development of the cut set is finished.

It can happen, however, that some repeated gates are substituted in the sequence proposed and reappear several times during the development of the same set. Each time that the substitution of a gate in a set is repeated, nonminimal sets are created. In order to avoid this, an additional copy of each set is created. However, the gates are not erased from the copy when they are developed. A gate is only added to a set during substitution if the gate does not exist in the copy already. The computational advantages of this algorithm compared with a normal top-down algorithm for enumerating cut sets have been shown in Reference 13.

V. SOME EXAMPLES

The fault tree presented in Figure 8 can contain more than 10^{17} cut sets despite its relatively small size. This tree cannot be decomposed in standard modules by the algorithm of section II.A, and a direct determination of its dominant cut sets for the components failure probabilities given in Reference 17 was impossible within reasonable computing time. Using the decomposition in supermodules shown in Figure 8, an estimated value of $2.18 \cdot 10^{-5}$ was obtained for the top event probability. The calculation was carried out without cutoff procedures, because the tree parts to be calculated separately were simple enough. This value might include the contribution of nonminimal cut sets. A demodularization process permits them to be eliminated and also to obtain the dominant MCSs of the system, which normally are desired information. Experience shows that the demodularization can be performed much faster by a top-down algorithm than a normal cut set enumeration of the tree, because the intermediate states of the development, which now are modular functions rather than gates, have been evaluated before and a cutoff criterion is highly efficient. The entire evaluation process described finds the 82 further relevant MCSs of the tree, which give a failure probability of $1.944 \cdot 10^{-5}$ considering the first term of the inclusion-exclusion method. The truncated part is estimated to be less than $1.26 \cdot 10^{-7}$. The computing time on a Cyber 170-835 was 3 sec.

A second fault tree for a part of a reactor protection system is presented in Reference 17. This new fault tree is larger than the former one and cannot be presented here for technical reasons. The tree contains more than 10^{30} cut sets, of which approximately $1.1 \cdot 10^{7}$ are minimal. The fault tree is so highly interlinked, that its modularization by standard procedures leads to an only slightly simplified tree which still contains more than 10^{26} cut sets, and cannot be evaluated within a reasonable period of time. However, if supermodules are not excluded from the algorithm, the most probable 272 MCSs are obtained in 41 sec., giving a top event probability of $1.0446 \cdot 10^{-8}$ and estimating the contribution of the neglected cut sets to be less than $1.097 \cdot 10^{-9}$. In the demodularization process, only 1.5 sec. of computing time were consumed. Eight supermodules of order between 3 and 11 were selected automatically, and a total of 219 products of modular functions were evaluated separately.

Other satisfactory results have been also obtained in safety studies from the international use of the code with large fault trees.

VI. CONCLUDING REMARKS

The modularization theory exposed is quite general and includes any other module concept for binary systems presented up to date. Nevertheless, attention was especially focused on coherent systems, for which the algorithms proposed were developed. It would, however, not be especially difficult to extend it to the evaluation of other types of systems. This algorithm was implemented on the computer code TREEMOD and its usefulness was proved in cases where standard modularization techniques do not give successful results.

A further improvement of the method for selecting supermodules, with the aim of optimizing the global efficiency of the evaluation process, can be a point for future work.

REFERENCES

1. **Vesely, W. E., Goldberg, F. F., Roberts, N. H., and Haasl, D. F.,** Fault tree handbook, NUREG-0492, Nuclear Regulatory Commission, Washington, D.C., 1981.
2. **Matthews, S. D.,** MOCARS: A Monte Carlo simulation code for determining distribution and simulation limits, ERDA Report TREE-1138, EG & G Idaho, Idaho Falls, Idaho.
3. **Camarinopoulos, L.,** Anwendung von Monte Carlo-Verfahren zur Ermittlung von Zuverlässigkeitsmerkmalen technischer Systeme, ILR-Bericht 14, TU Berlin, 1976.
4. **Fussell, J. B. and Veseley, W. E.,** A new methodology for obtaining cut sets from fault trees, *Trans. Am. Nucl. Soc.,* 15, 262, 1972.
5. **Chatterjee, P.,** Fault Tree Analysis: reliability theory and system safety analysis, ORC 74-34 Operations Research Center, University of California.
6. **Rosenthal, A.,** A computer scientist looks at reliability computations, in *Reliability and fault tree analysis,* SIAM, Philadelphia, Pa., 1975, 133.
7. **Birnbaum, Z. W. and Esary, J. D.,** Modules of coherent binary systems, *J. Soc. Ind. Appl. Math.* 13(2), 444, 1965.
8. **Shapley, L. S.,** On committees, in *New methods of thought and procedure,* Zwycky, F. and Wilson, A., Ed., Springer, N.Y., 1968, 246.
9. **Billera, L. J.,** On the composition and decomposition of clutters, J. *Combinatorial Theor.* 11, 234, 1971.
10. **Billera, L. J.,** Clutter decomposition and monotonic Boolean functions, *Ann. N.Y. Acad. Sci.,* 175, 41, 1970.
11. **Chatterjee, P.,** Modularization of fault trees: a method to reduce the cost of the analysis, in Reliability and fault tree analysis, SIAM, Philadelphia, Pa., 1975, 101.
12. **Olmos, J. and Wolf, L.,** A modular approach to fault tree and reliability analysis. MITNE-209, Cambridge, Mass., 1977.
13. **Camarinopoulos, L. and Yllera, J.,** An improved top-down algorithm combined with modularization as a highly efficient method for fault tree analysis, *Reliab. Eng.* 11(2), 93, 1985.
14. **Chu, T. L. and Apostolakis, G.,** Methods for probabilistic analysis of noncoherent fault trees. *IEEE Trans. Reliab.,* 29(5), 354, 1980.
15. **Caldarola, L.,** Coherent systems with multistate components, *Nucl. Eng. Des.,* 58, 127, 1980.
16. **Yllera, J.,** Modularisierungsverfahren zur Berechnung von Fehlerbäumen komplexer technischer Systeme, Doctoral Thesis, Technische Universität, Berlin, 1986, DS83.
17. **Caldarola, L. and Wickenhäuser, A.,** The Boolean algebra with restricted variables as a tool for fault tree modularization, KfK 3190, EUR 7056e.
18. RISA Program Description, Version 4., CD Manual, September 1983.
19. **Aho, A. V., Hoptcroft, J. E., and Ullman, J. D.,** *The Design and Analysis of Computer Algorithms,* Addison-Wesley, Reading, Mass., 1975, 171.

Chapter 6

UTILIZING PROBABILISTIC RISK ANALYSES (PRAS) IN DECISION SUPPORT SYSTEMS

William E. Vesely

TABLE OF CONTENTS

I. INTRODUCTION

One of the most useful applications of probabilistic risk analysis (PRA) is to use the PRA to identify the risk importances of design features, plant operations, and other factors that can affect risk. Risk importance as it is commonly used in PRA terminology, is generally the impact on risk that a factor has. The definition and concept of risk importance will be expanded as this discussion unfolds. Utilization of risk importances in decision support systems utilizing microcomputers and minicomputers will be of particular interest here. The material in this discussion is adapted from the report, by Vesely and Davis.[1] This discussion focuses on nuclear power plant applications, but it is applicable to general PRA utilizations.

PRAs can be used to identify the importances of risk contributors and proposed changes to systems and operations. The risk contributors can include system failures, component failures, human errors, and test and maintenance deficiencies that contribute to risk. The particular risk which is focused upon can be a safety system unavailability, core melt frequency, expected latent fatalities, or various other risk measures. The risk and risk contributors which are examined will depend upon the specific objectives of the application.

PRAs can also be used to evaluate the risk importances of changes in designs, operations, or plant conditions. These changes can be either beneficial changes, such as design improvements, or can be deleterious changes, such as component wearout. The risk importance is then the impact or change in risk resulting from the input change. The particular changes which are evaluated will again depend upon the objectives of the PRA utilizations.

The risk importances which shall be focused upon here are those importances which are derived from the logic structure of the PRA. The logic structure of the PRA includes the fault tree and event tree models, the failure combinations causing undesired events (the minimal cut sets), and the success paths preventing undesired events (the minimal path sets). The structural and logical information in a PRA provides valuable criteria by which to evaluate importances of risk contributors and changes. The importances of risk contributors and changes can be evaluated with regard to the estimated system failure probabilities, accident sequence frequencies, core melt frequency, or the accident frequency vs. consequence curve.

II. DISTINCTION BETWEEN RISK REDUCTION AND SAFETY ASSURANCE

Before discussing the evaluation of importances, it is useful to first identify the general uses of importances. In using risk importances, one can have either one of two general objectives:

1. Risk reduction.
2. Safety assurance.

The objective of risk reduction is to make the present risk lower. This implies that some decision has to first be made that the present risk level is unacceptable. Even if cost benefit analysis is used to show a positive net benefit from the risk reduction, in order to carry out any action there still has to be a decision made that the present risk level is unacceptable because a positive cost benefit can be shown for a risk reduction. Examples of risk reduction activities which have been carried out are present plant backfitting programs and the plant changes that were instituted after the Brown's Ferry fire and the Three Mile Island accident.

In evaluating risk importances for risk reduction, the focus is on identifying the dominant contributors to present risk. Ways for reducing these contributors are then assessed for their cost, effectiveness, and other criteria.

The objective of safety assurance, or reliability assurance as it is sometimes called, is to assure that risk does not increase and is as low as the PRA indicates it is. There is no determination, as in the risk reduction case, that risk should be reduced because it is unacceptable or provided a positive net benefit can be demonstrated. The prime objective of safety assurance is to protect against a risk increase. Examples of safety assurance activities are plant inspection activities performed by NRC inspectors and reliability assurance programs carried out by the plant personnel.

When evaluating risk importances for safety assurance uses, the focus is on those conditions which have the greatest impact, or importance, in increasing risk. The importance of a safety assurance activity or safety assurance feature is consequently the impact it has in keeping the risk from increasing.

In a plant, risk reduction and risk assurance activities can both be carried out concurrently. If a risk assurance activity finds an increased risk condition, then the situation and objective change. If the increased risk is assessed to be significant, then the objective now becomes one of risk reduction. Similarly, a risk reduction activity, once a risk reduction change is instituted, will transform to safety assurance activities to assure the change is effectively instituted and is maintained.

Plant activities and functions can thus have both risk reduction and risk assurance aspects. It is useful however, to separate the risk reduction and safety assurance objectives. This is particularly useful for risk importance evaluations since the type of risk importance which is applicable will depend not only on the general objective, but also on the specific problem being addressed.

III. EVALUATION OF RISK IMPORTANCES

As was stated, risk importances are importances derived from the logic structures of the PRA. The logic structures which can be specifically examined for their importance information are

1. The event tree and system models themselves
2. The critical failure combinations, or minimal cut sets, of the PRA
3. The failure combinations which are not critical but will cause significant risk increases
4. The success paths, or minimal path sets, of the PRA

Utilization of these logic structures to derive risk importances will be described in the following sections. As part of the discussions, applications will be presented with specific attention paid to utilizations of PRAs as decision support tools. Micro/minicomputers provide the means for the decision maker to interact on a real time basis with the PRAs as decision support tools. The basic approaches will first be described followed by examples of decision support systems being developed which utilize the risk importance information.

A. The Event Tree and System Models

The event trees and fault trees developed in the PRA are the first sources of information on what is important to risk. The event trees and fault trees are logic representations of what is necessary to cause undesired events to occur. The event trees define the specific sequences of failures that are necessary for a core melt or a radioactive release to occur. The failures in the accident sequences can be safety function failures or safety system failures, depending upon the resolution of the event trees. The fault trees define the specific component failures that are necessary for a particular safety function or safety system to fail. The construction of fault trees and event trees for risk analyses is described in References 2 and 3.

The problem with the original event trees and fault trees in the PRA is that they are cumbersome and are often expressed in PRA jargon. The original event trees and fault trees were developed as base models for the purpose of evaluating all potential contributors to risk that were postulated. These models were not generally edited after the analysis was performed to make them utilisable in other applications.

The PRA models can be edited and accessed in several ways:

1.   Show only the dominant contributors
2.   Present the models in a hierarchical way
3.   Place the models on micro/minicomputers for easy access

Keeping only the dominant contributors reduces the models significantly; at times, over 75% of the contributors are not risk significant. Care must be taken in extracting the dominant contributors, since what is dominant will depend upon the application. This is generally not a problem, since the contributors can be straightforwardly organized according to different applications and objectives. For example, for accident prevention, the dominant contributors to core melt frequency and to system unavailability would be the focus. For consequence mitigation, the dominant contibutors to health consequences, given a core melt, would be the focus.

Arranging the models in a more hierarchical way than is presented in the PRA can greatly organize the information to make it more accessible and understandable. Accident sequences leading to core melt or core damage can be presented in terms of the initiating event and failures of safety functions which are required. Failures of safety functions can be presented in terms of failures of safety systems which cause the function to be unavailable. Finally, failures of safety systems can be presented in terms of necessary train failures and component failures, In this way, the user can step through the risk contributors to the level and detail desired.

As an efficient way of accessing the PRAs, the models and evaluations can be placed on a micro/minicomputers. The plant operator or inspector can then simply query the models to obtain risk information. Having the models on a micro/minicomputer is particularly useful for identifying risk information, such as dominant contributors, as a function of a particular plant status.

The PRA jargon problem can be addressed by describing events in English. Even if acronyms and abbreviated names are used, they can be suggestive of the events. Effective display and communication of the PRA models is not given the attention it should in PRAs. Micro/minicomputer utilizations can also be helpful with user menus and help commands.

The event tree and fault tree models, in edited and accessible form, can provide a key tool for educating plant personnel and inspectors on how to think of the plant in terms of risk. This knowledge can be used as a foundation for how the personnel then perceive events and issues in terms of their risk significance.

The displays on the following pages provide illustrations of the ways PRA information can be presented on a minicomputer. The displays were taken from a presentation developed from the PRA on the Arkansas Nuclear One (ANO) plant.[2] The displays are self-explanatory and can be further enhanced if desired, using color, multiple windows, and further graphical treatments. The individual displays can be accessed via a menu, or a sequence of displays may be viewed as a tutorial.

B. The Critical Failure Combinations, or Minimal Cut Sets

A critical failure combination, or minimal cut set as it is called in PRA terminology, is a smallest combination of component failures that will result in an undesired event. The undesired event can be a system failure, function failure, or an accident sequence occurrence.

Consider the simple series system below. The system could represent, for example, a train of a safety system. This simple system has three single component minimal cut sets or min cut sets for short. This simply says that if any of the components fail, then the system fails.

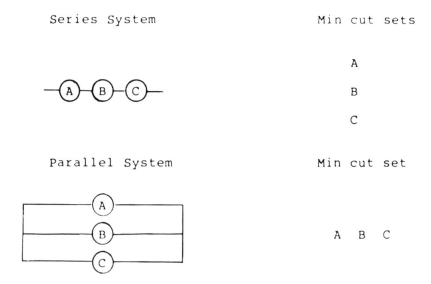

Series System   Min cut sets

A

B

C

Parallel System   Min cut set

A   B   C

For the parallel system above, there is one three-component min cut set, which says that all three components must fail for the system to fail.

The min cut sets for a general system failure are all those combinations of component failures that result in system failure. Each component failure combination is unique and is minimum, in that all the component failures in the combination are necessary to cause system failure in this particular way. An example of a system min cut set is the failure of two electric pumps and a steam driven pump which fail the auxiliary feedwater system in a nuclear power plant; the min cut set in this case consists of three component failures.

The min cut sets of a safety function are the combinations of component failures which result in the safety function failure. The min cut sets of an accident sequence are the component failure combinations which result in the accident sequence occurring.

The PRA provides the min cut sets of each system failure, each function failure, and each accident sequence occurrence. There are only a finite number of min cut sets for each undesired event (system failure, function failure, etc.). The component failures can be actual component hardware failures (e.g., a pump failure) or can be human errors or operational causes of failures ("component" is thus used in a general sense here). The min cut sets are usually ordered with regard to the number of components in the min cut set and the min cut sets are usually truncated after the number of component failures in the min cut set exceeds some value. Min cut sets are described in more detail in the Fault Tree Handbook[2] and in the PRA Procedures Guide.[3]

The min cut sets are the key quantities used in quantifying the PRA. The min cut sets provide structural information which can be used to identify important component failures and important situations that can lead to high risks. Single component min cut sets for a system, for example, are single component failures which result in system failure. The Nuclear Regulatory Commission's (NRC) single failure criterion does not allow particular types of single failures in nuclear plant safety systems and the min cut sets can be checked to assure that there are no single cut sets.

One of the uses of min cut sets that can be particularly applicable to inspection and reliability assurance is to identify common cause failure conditions which can significantly

increase risk and which, therefore, must be guarded against. Based on reliability considerations, it is known that components which have a common failure susceptibility are subject
to failure from one common cause which can fail all the components. Examples of components having a common failure susceptibility are given below.

Components having a common susceptibility	The common susceptibility	The common cause which can trigger all the failures
Components in a common location	The common location	Fire or another ener- getic event in the location
Similar components under the common maintenance	The common maintenance	An error in the maintenance

PRAs tell us something very significant about what common cause susceptibilities to focus
on. From the PRA min cut set definition, the common susceptibilities are risk significant,
or risk important, only if the susceptible components are in the same min cut set. If the
susceptible components are in the same min cut set, then a common cause can trigger the
component failures, which because they are in the same min cut set will then trigger a system
failure or accident sequence. If the susceptible failures are not in the same min cut set, then
no immediate consequences will result from the common cause triggering of the failures.
Other independent failures are required, which thereby causes these contributions to be lower
in probability and, hence, less risk significant.

From the PRAs that have been performed to date, common cause failures of susceptible
components in the same min cut set are generally dominant contributors to risk. The min
cut set information from PRAs can be used in following the stepwise approach in inspection
and safety assurance activities to guard against common cause failures:

1.   Identify those min cut sets where all the components have a common susceptibility.
2.   From the components in the susceptible min cut sets, assemble a checklist of these
     components for the inspector and for safety assurance programs. Components in the
     same min cut set can be identified as critical groups. This list can be labeled as a
     common cause failure checklist.
3.   Whenever one of the check-listed components have been found failed or degraded,
     the inspector can assure or have the plant assure that the other components in the
     critical group (min cut set) have not been affected by the failure cause. This assurance
     is especially critical if failures of the other possible components are not readily detectable.
4.   To further strengthen assurance against these common cause failure potentials, specific
     protections can be developed and be documented for each of the check-listed critical
     groups.

The above approach can be used to focus on and to give substance to safety assurance
activities directed to guarding against common cause failures. The min cut sets provide
specific components and susceptibilities which should be guarded against because of their
risk significance. The construction of a checklist simplifies implementation with knowledge
of PRAs not really necessary to use the checklist. The instructions can be described in clear,
understandable language, which is not tied to PRA jargon. Furthermore, if the checklists
are programmed on a micro/minicomputer, they can be readily accessed to provide specific
information for any given situation.

The common susceptibilities in a min cut set which can be specifically focused upon
include:

1. All components of the same generic type (such as all pumps) in a min cut set (critical group) indicative of potentially common, critical vulnerabilities.
2. All components in a min cut set in the same location.
3. All components in a min cut set under the same maintenance or testing procedure.
4. All human errors in a min cut set which implies that human errors alone can fail critical subsystems, systems, or functions.
5. All components in a min cut set which can be exposed to a common harsh or degrading environment.
6. All components in a min cut set not testable in routine surveillance testing, thereby giving a critical undetectable failure mode.
7. All components in a min cut set tested under a common pre-op or start up procedure: if the procedure is inadequate then a critical failure mode can be untested or undetected.

The level of the min cut sets, e.g., min cut sets at the system level, determine the level at which the safety assurance against common cause failures is focused. The most comprehensive approach is to address safety assurance at various levels (system, function, and accident sequence) to provide a multi-level, multi-pronged protection. Documentation should be performed to validate that assurance has been taken against the identified common cause susceptibilities. Applications of this information in decision support systems will be deferred until other risk importance information is described.

C. Vulnerable Plant States

The min cut sets can also be used to identify plant conditions during which the plant is at higher risk and is more vulnerable to accident occurrences. From a safety assurance standpoint, it is important that these high risk conditions first of all be recognized. When these conditions occur, extra precaution can then be taken and efforts can be focused at moving the plant from this high risk state. Furthermore, where feasible, specific procedures can be instituted to avoid these high-risk situations. The role of the inspector or plant operator can be particularly important in assuring that these high-risk conditions are recognized and protected against.

The min cut sets are useful for identifying these important, high-risk situations. When a component or components fail such that there is only one remaining unfailed component in a min cut set, then protection against the undesired event is riding on that component. If that component were to fail, then undesired consequences will occur. The situation will be an especially higher risk situation if the one remaining unfailed component has a higher unreliability. Active components (pumps, valves, etc.) and human errors generally have these high unreliabilities.

The min cut sets can be used in the following useful ways to identify specific high-risk situations:

1. Identify all single failures, or single components being down, which result in one active component or one human error remaining in a min cut set. Identify also the active component or human action which is providing the sole protection against the undesired event.
2. Assemble the above single failures in a checklist which also identifies the remaining protection.
3. Identify those double failures, or double components being down, which result in one active component or one human error remaining in a min cut set. Identify also the remaining component or human action providing the sole protection against the undesired event.
4. Assemble the above double failures and remaining protection in a checklist which identifies the high-risk situations caused by two components being down.

The above checklists can be used in a straightforward manner in inspection and in safety assurance programs. The checklists provide specific high-risk situations to be aware of, and an attitude of extra vigilance should prevail during these high-risk periods. Where feasible, specific activities can also be identified which provide added assurance for the remaining component or human action providing the sole protection. For the high risk situations caused by two components being down, procedures can also be identified for avoiding these high-risk situations, which for example can be caused by multiple components being simultaneously brought down for maintenance without an awareness of the risk implications. If the checklists are programmed on a micro/minicomputer, they can be readily accessed to show those components which should not be down at the same time and those critical components providing sole protection when a component or components are down for a given repair or maintenance.

D. The Success Paths, or Minimal Path Sets

The success paths, or min path sets as they are called in PRA terminology, are another type of structural information provided by PRAs.* The min path sets are the complements of the min cut sets and are the ways the undesired event can be prevented.

A min path set is a smallest combination of components which if assured to all be up will assure that the undesired event will not occur. For a system, the system min path sets are all the unique, minimal ways that the system can be assured to be up. The min path sets of an accident sequence are all the unique, minimal ways that the accident sequence can be prevented.

By the min path set definition, only one min path set needs to be assured to be up to assure against the undesired event. A min path set being assured is again assuring that all components in the min path set are up. If the "component" is a human action, then the assurance constitutes assuring a successful human action; If the "component" is a plant condition, then the assurance constitutes assuring the plant condition does not exist. If the assurance is a test, then each min path set gives the combination of component tests which are necessary for an integral test to be performed on the system, of the safety function, or for prevention of an accident sequence.

The level of the min path sets (e.g., at the system level) determines the level at which overall safety assurance or integral testing is focused. PRAs can provide the min path sets for each system, each function, and each accident sequence. The min path sets can be organized and can be categorized using various criteria. The min path sets have not been standardly calculated in PRAs. This should change, however, with more attention being directed to the use of PRAs for safety assurance and reliability assurance activities. The actual determination of the min path sets is straightforward and involves standard Boolean operations.

To illustrate the min path set concept, consider the simple example shown below. For the series system, there is one min path set consisting of all the components. This simply says that to assure a series system is up, such as a train of a safety system, all the components must be assured to be up.

For the parallel system, there are three single component min path sets, since if any component is assured to be up, the system is assured to be up.

For complex systems and accident sequences, the min path sets are not obvious and the PRAs provide this important information. The min path sets are key to an understanding of what constitutes an overall assurance against the undesired event or what constitutes an integral test. The min path sets can provide concrete information which can be used to focus and organize specific safety assurance activities and programs.

---

*   The min path sets are also sometimes called protection sets, also an appropriate term.

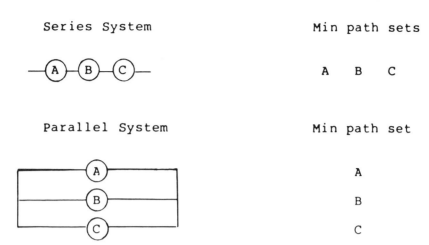

The min path sets can be used in inspection and in safety assurance in the following specific ways. First of all, there needs to be a recognition that to assure against an undesired event, at least one min path set (success path) for that event needs to be assured. This concept is basic and can be included in education and orientation programs.

Using the min path set concepts, the following specific questions can then be asked in inspection and safety assurance activities with answers being provided by the min path sets:

1. Are there any min path sets (success paths) which can be assured by routine testing or monitoring? If so, these important testable success paths can be identified in a checklist. The component tests constituting a success path can then be inspected and be assured as an integral group.

2. If there are no success paths which can be assured solely by routine testing or monitoring, then to what degree is an integral test or overall assurance possible? This can be answered by identifying those key components which are not testable, but are necessary for an overall success path assurance.

3. With regard to the key components, which are not testable by routine testing or monitoring, can success path assurance be provided by alternate means such as testing at shutdown, testing at pre-op or start up, equipment qualification tests, or plant demands (transients)? If so, the specific means or events should be identified as being key to overall success path assurance. Inspection and quality assurance can focus on these key activities or events.

4. If there are no success paths which can be assured by any type of direct testing (or plant demands), then what indirect assurance is needed for an overall success path assurance? This indirect assurance can include, e.g., calculational techniques or sampling techniques. This indirect assurance can be identified as being critical for overall success path assistance and can also be the focus of inspections and reviews.

The above approaches can be modified or can be tailored in various ways. The above approaches focus on the a issue in safety assurance which is the identification of success paths and what constitutes assurance of a success path. Attention is systematically directed towards those success paths which can be effectively tested or be assured, and to those components which are key to integral testing and overall assurance. The above approaches can be useful, even if they're only used to help organize and structure present safety assurance activities. If programmed on a micro/minicomputer, the success paths can be readily accessed for any plant status, providing timely information on what success paths are now available.

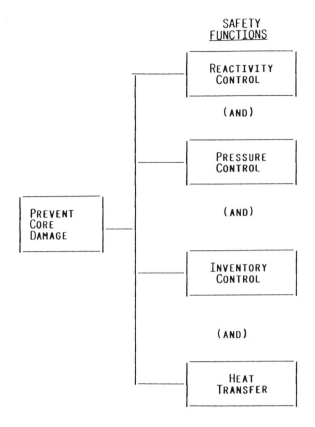

FIGURE 1.   Display of required safety functions at the Arkansas
Nuclear One plant.

E. Demonstration of Importance Utilizations

Figures 1 to 5 on the following pages demonstrate utilization of importances discussed in the previous sections. The displays give safety assurance checklists which can be utilized by the NRC inspector or plant personnel. The PRA for Arkansas Nuclear One (ANO)[4] is again used as the information source. The Emergency Feedwater System is specifically illustrated as an example because of its large impact on core melt frequency. (Similar displays are obtainable for all the systems in the plant.)

In actual utilization, the NRC inspector or the plant operator selects the desired information category, the system, and the component(s) of interest on a simple program menu. The appropriate displays are then shown giving the information and recommended actions.

The displays are samples and are fairly self-explanatory. Figure 6 identifies those critical groups (min cut sets) of components which are susceptible to common cause failures because of the similarity of components. Figure 7 identifies those specific components to check when a given component has failed: on a micro/minicomputer, such a checklist can be quickly accessed to focus inspection and safety assurance when an incident has occurred.

Figure 8 is a display of an operability checklist based on vulnerable plant states. Again, such a checklist can provide timely and critical information on key components to assure operation when a given component is down for maintenance. Figure 9 is a display of what assurances or checks are needed to provide integral assurance that the emergency feedwater system is up; Figure 10 is a graphic display of integral assurance requirements. The integral assurance requirements are based on the success paths.

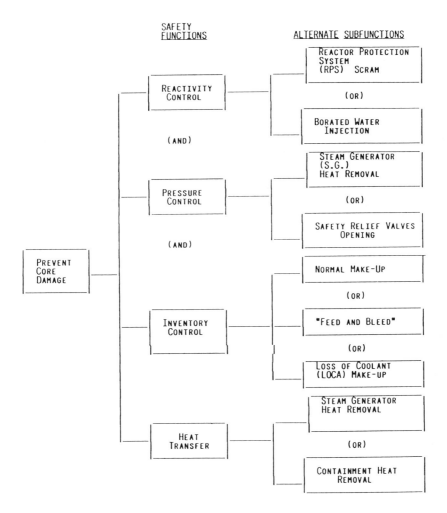

FIGURE 2.   Display of alternate subfunctions for achieving the required safety functions at Arkansas Nuclear.

IV. SUMMARY

PRAs provide a comprehensive information base on which to build decision support systems to assist both the NRC personnel and the plant personnel in performing their functions. Risk importance information in particular provides useful information to guide their decisions and actions. The previous sections have provided an overview of the specific ways in which risk importance information can be utilized to build decision support systems on micro/minicomputers. The previous sections, however, provide only the tip of the iceberg of the ways PRAs can be utilized in decision support systems and, with further extensions, in expert systems. Currently, there is a significant amount of work underway in the nuclear community in developing these systems, and in the upcoming future there will be a great expansion of this area.

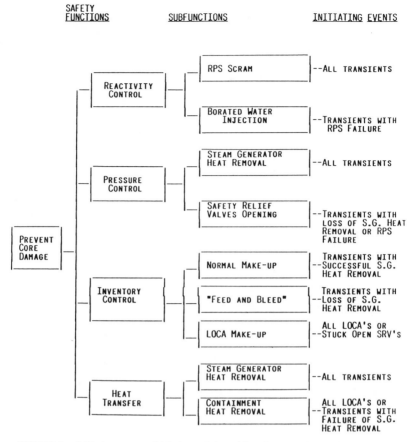

FIGURE 3.   Initiating events which demand the subfunctions at Arkansas Nuclear.

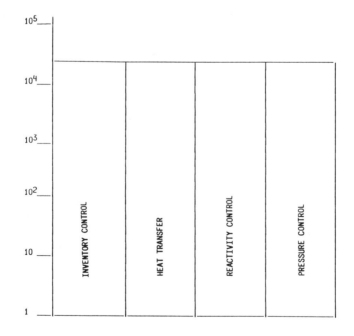

FIGURE 4.   Factor by which the core melt frequency would increase if
a particular safety function were unavailable.

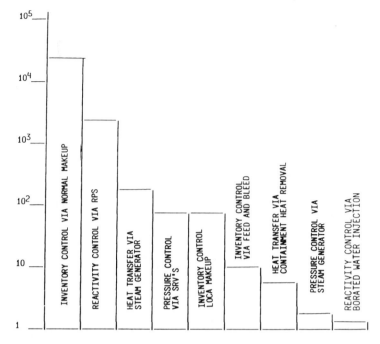

FIGURE 5. Factor by which the core melt frequency would increase if particular subfunctions were unavailable.

TITLE: COMMON CAUSE CHECKLIST FOR EMERGENCY FEEDWATER SYSTEM

INFORMATION   Groups of components which are susceptible to the same
CATEGORY: failure cause and which if all fail will fail EMERGENCY
                    FEEDWATER SYSTEM

DESIRABLE   Assure that components in each group are protected from
ASSURANCE   common or systematic failure causes especially common
ACTION: maintenance or testing errors and common environmental
                    stresses

SPECIFIC GROUPS OF COMPONENTS AND FAILURE MODES

GROUP 1.   MOTOR DRIVEN PUMP 1 (INOPERABLE)
           MOTOR DRIVEN PUMP 2 (INOPERABLE)

GROUP 2.   MOTOR OPERATED VALVE 1 (FAILED CLOSED)
           MOTOR OPERATED VALVE 2 (FAILED CLOSED)

GROUP 3.   MOTOR OPERATED VALVE 3 (FAILED CLOSED)
           MOTOR OPERATED VALVE 4 (FAILED CLOSED)

GROUP 4.   CIRCUIT BREAKER 3 (FAILED OPEN)
           CIRCUIT BREAKER 4 (FAILED OPEN)

FIGURE 6. Display of a common cause failure checklist for Arkansas Nuclear.

```
TITLE:              COMMON CAUSE CHECKLIST WHEN A COMPONENT IS FOUND FAILED

INFORMATION         Components which are susceptible to a common cause failure
CATEGORY:           and if down will cause significant risk increases

DESIRABLE           Assure  that  components  in  each  group  have  not  been
ASSURANCE           affected by the failure of the given component
ACTION:

QUERY:              Failed component?   Motor  Operated  Valve 1 System ?
                    EMERGENCY FEEDWATER


        OTHER SUSCEPTIBLE COMPONENTS              CONSEQUENCES OF THESE
                TO CHECK                          ADDITIONAL FAILURES


        GROUP 1.   MOTOR OPERATED VALVE 2         EMERGENCY FEEDWATER
                   (FAILED CLOSED)                    FAILED

        GROUP 2.   MOTOR OPERATED VALVE 4         EMERGENCY FEEDWATER
                   (FAILED CLOSED)                    FAILED
                       (AND)
                   MOTOR OPERATED VALVE 5
                   (FAILED CLOSED)

        GROUP 3.   MOTOR OPERATED VALVE 6         EMERGENCY FEEDWATER
                   (FAILED CLOSED)                    FAILED
                       (AND)
                   MOTOR OPERATED VALVE 7
                   (FAILED CLOSED)



IF EMERGENCY FEEDWATER IS FAILED, CORE MELT FREQUENCY INCREASES BY A
                    FACTOR OF 70
```

FIGURE 7.   Display of a common cause checklist for a failed component at Arkansas Nuclear.

```
TITLE:              OPERABILITY CHECKLIST WHEN A GIVEN COMPONENET IS DOWN FOR
                    REPAIR OR MAINTENANCE

INFORMATION         Critical  components  which  if  also  down  will  cause
CATEGORY:           significant risk increases

DESIRABLE           Assure that other critical components are up and operable
ASSURANCE           during the repair or maintenance
ACTION:

QUERY:              Downed component?  Motor Driven PUMP 1 System?  EMERGENCY
                    FEEDWATER


          COMPONENTS TO CHECK                  CONSEQUENCES IF COMPONENTS
                                                     ARE  DOWN


        1.   MOTOR DRIVEN PUMP 2               EMERGENCY FEEDWATER
             (OPERABLE)                            FAILED

        2.   TURBINE PUMP                      EMERGENCY FEEDWATER
             (OPERABLE)                            FAILED

        3.   TURBINE CONTROL VALVE             EMERGENCY FEEDWATER
             (OPEN)                                FAILED

        4.   MOTOR OPERATED VALVE 2            EMERGENCY FEEDWATER
             (ABLE TO OPEN)                        FAILED


IF EMERGENCY FEEDWATER IS FAILED, CORE MELT FREQUENCY INCREASES BY A
                    FACTOR OF 70
```

FIGURE 8.   Display of operability checklist when a component is taken down at Arkansas Nuclear.

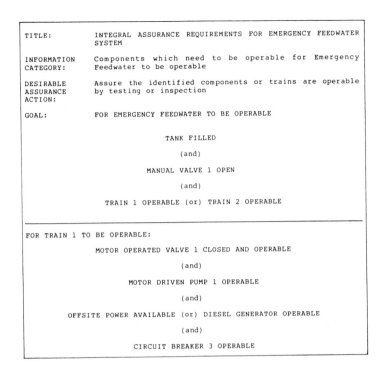

FIGURE 9.   Display of integral assurance requirements at Arkansas Nuclear.

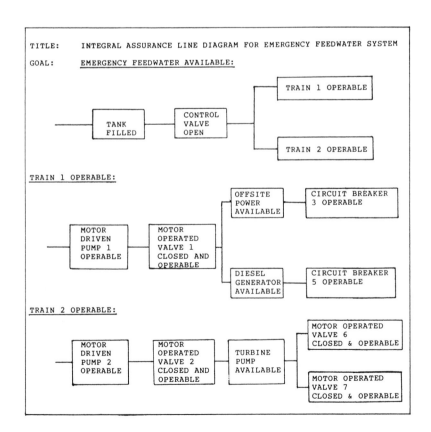

FIGURE 10.   Graphic display of integral assurance requirements at Arkansas Nuclear.

REFERENCES

1. **Vesely, W. E. and Davis, T. C.,** *Evaluations and Utilization of Risk Importance,* NUREG/CR-4377, Nuclear Regulatory Commission, Washington, D.C., 1985.
2. **Vesely, W. E., Goldberg, F. F., Roberts, N. H., and Haasl, D. F.,** *Fault Tree Handbook* NUREG-0492, Nuclear Regulatory Commission, Washington, D.C., 1981.
3. American Nuclear Society and the Institute of Electrical and Electronics Engineers, *PRA Procedures Guide,* NUREG/CR-2300, Rev. 1, Nuclear Regulatory Commission, Washington, D.C., 1982.
4. **Kolb, G. J.,** *Interim Reliability Evaluation Program: Analysis of the Arkansas Nuclear One-Unit 1 Nuclear Power Plant,* NUREG/CR-2787, Nuclear Regulatory Commission, Washington, D.C., 1982.

Chapter 7

SURVEY OF EXPERT SYSTEMS FOR FAULT DETECTION, TEST GENERATION, AND MAINTENANCE

L. F. Pau

TABLE OF CONTENTS

I. INTRODUCTION

A. Knowledge-Based Systems

To improve the computerizing of failure detection, testing, and maintenance, artificial intelligence and knowledge based techniques are currently being explored.[7,8,21,22,28,35] Knowledge-based systems (KBS) are software programs supplemented with man-machine interfaces, which use knowledge and reasoning to perform complex tasks at a level of performance usually associated with an expert of either of these three domains.

An expert system essentially consists of a knowledge base containing facts, rules, heuristics, and procedural knowledge, and an inference engine which consists of reasoning or problem solving strategies on how to use knowledge to make decisions.[7,28,36] It also consists of a user interface with the user in either natural language, via interactive graphics, or through voice input. The explanation generator in the expert system provides answers to queries made by the user. The knowledge base is developed by a small group of knowledge engineers who query the domain expert(s). As an aid to getting knowledge into the knowledge base, a knowledge acquisition tool is used (either by the domain expert or knowledge engineer).

B. Applications to Failure Detection, Testing, and Maintenance

Failure detection, testing, and maintenance are knowledge intensive and experience-based tasks.[7,8,21] Although test procedures and maintenance manuals contain recommended detection, localization, testing, maintenance, and repair actions, their use alone does not assure successful completion of trouble shooting and repair in a timely manner. Skilled maintenance staff, apart from using test procedures and maintenance manuals, use heuristics and an understanding of how the system works to solve problems. It is this ''beyond procedures'' type of knowledge that enables them to perform at an exceptional level. Based on years of experience, a highly skilled test/maintenance technician develops the following traits:

1. A familiarity with procedures and documented maintenance manuals
2. An understanding of least repairable unit (LRU) and symptoms interactions
3. An understanding of the relationships between symptoms and failed LRUs
4. An intuitive understanding of how the system works
5. An intuitive understanding of how the system will behave when certain subsystems or LRUs fail

The high level of performance of experts suggests that, confronted with a problem, they analyze the problem in a structured manner rather than randomly trying all possible alternatives. Experience with medical diagnosticians, in particular, suggests that expert diagnosticians have their diagnostic knowledge organized in powerful hierarchical structures that enables them to quickly reason from given symptoms to specific system level problems to LRU level problems, using various testing procedures wherever appropriate.

Problems with an aircraft are typically reported through pilot squawks.[21] Pilot squawks contain information on lost capabilities of aircraft functions. Based on the reported problem, the pilot is debriefed for more specific information, during which the test/maintenance specialist tries to narrow the list of possible malfunctioned LRUs. Sometimes the malfunctioned LRU can be identified based on the debriefing session. During the debriefing session, the specialist is *interpreting* the symptoms and *diagnosing* the problem by asking more specific data. Often at the end of the debriefing, the specialist will have limited the malfunctions to a few LRUs. The specialist will then *troubleshoot* if the failed LRU is identified, and then the appropriate *replacement/repair* action will be taken. After the repair action is complete, the system is retested and the response *monitored* to assure that the problem is removed.

Sometimes the problem cannot be easily diagnosed. In such situations, the historical database of the specific aircarft and the fleetwide database are consulted in order to obtain a clue.[21] Failing this, the test/maintenance specialist has to use his *deep understanding* (i.e., the knowledge on how the system works) to diagnose the problem and sometimes *design* new tests to test for unusual conditions. Finally, the logical and structural coherence of symptoms, tests, and maintenance actions must be checked for final detection decisions, test selection, and repair. This decision level quite often involves pattern recognition techniques.[21] Because this process is time consuming, special considerations may be required under wartime conditions. Due to the constraints of limited resources (available technicians, their skill level, available spare parts, and testing equipment) and available time for aircraft repair, the tests to be performed, and the repair actions to be taken are *scheduled* in order to effect a short turn-around time.

In designing a KBS that will perform the above mentioned tasks, several types of tasks will need to be modeled. Identifying these tasks and associated reasoning processes distinctly, and modeling them as independent modules is extremely important to achieving a high degree of performance and modularity for future expansion and modification of the system. Some of the tasks identified in the above discussion are interpretation, diagnosis, troubleshooting, repair, design, planning and scheduling, monitoring, reasoning with functional models, and metarules.[21,22,24]

II. EXAMPLES OF KNOWLEDGE-BASED FAULT DETECTION, TEST GENERATION, AND MAINTENANCE KBS

Several knowledge-based systems (KBS) exist that address some of the problems relevant to the detection/test/diagnosis/maintenance task. A review of the capabilities and limitations of some of these KBS is presented to identify methods and techniques that can be used. Other KBS exist or are under development, but cannot be mentioned here for lack of space and/or published open descriptions. KBS characteristics used here are defined in later sections.

MYCIN — MYCIN is one of the earliest KBS which diagnoses bacterial infections in blood samples. MYCIN was designed as an experimental system using a production systems approach. It uses backward chaining inferencing and has a modest size knowledge base. MYCIN contains around 700 rules. MYCIN solves only one type of problem-solving type: diagnosis. MYCIN can combine confidence factors associated with individual rules to obtain an overall confidence factor for the decision made. MYCIN's query facility is rather simple. It can inform the user why it is asking for certain data and how a certain decision was made. Designers of MYCIN have now decided to implement a similar system using a distributed problem solving approach.[27] MYCIN has under various names found its way into KBS shells, e.g., EMYCIN, several of which have been tested for diagnostics. GUIDON is a MYCIN-like program for teaching diagnosis,[7] and STEAMER[18] another one. However, MICYN, EMYCIN and the like suffer from the fact that diagnostic rules are tightly coupled to domain specific facts.

Failure detection schemes — These use pattern recognition techniques and corresponding learning information, and have been developed, e.g., for aircraft/missile engine monitoring, rotating machinery, and guidance systems.[21,24,26] Detection performances, as well as diagnostic results have often been excellent, provided failure modes could be characterized well enough from measurements and observations. None, however, includes explicitly symbolic knowledge, apart from implicit knowledge in the form of measurement data organization and logic conditions applicable to a hierarchical tree organization of the classification rules.

DELTA/CATS-1[7] — This is a production system type expert system that performs troubleshooting for locomotive engines. This expert system is a feasibility type demonstration system. The system was initially designed with 50 rules, and at last report had 530 such rules. Future plans are to expand the knowledge base. Like MYCIN, this system solves a single type of problem: diagnosis.

MDX — This is a distributed problem solving type KBS based on the notion of a society of specialists organized in a strictly hierarchical structure.[13] Apart from the "diagnosis" part, MDX also consists of other KBSs called PATREC and RADEX. The diagnosis portion interacts with these two auxiliary expert systems to obtain and interpret historical and lab data.

AUTOMECH — This is a KBS written in the CSRL language which diagnoses automobile fuel systems.[7] It is patterned after MDX.[13]

DART (used by IBM) — This is a diagnostic expert system that uses functional models of the components instead of diagnostic rules to diagnose a problem. DART is one of the earlier systems to use deep functional knowledge in designing expert systems.[16] Designers[14,33,39] have reported the use of deep knowledge in terms of system structure and behavior for diagnosing computers.

ISIS — This is a distributed problem solving type expert system designed to perform a job shop scheduling task in a manufacturing plant. In order to do job shop scheduling, ISIS has to take into account various types of constraints. Constraints considered by ISIS are (1) organizational constraints to assure profitablility (2) physical constraints to check the capability of a machine (3) gating constraints to check if a particular machine or tool can be used, and (4) preference constraints to enable the shop supervisor to override the expert program.[15] ISIS is a feasibility type demonstration system. It is still under development.

Diagnostic and test selection KBS shell — This has been developed and used for integrated circuit testing, datacommunications monitoring, avionics maintenance training, and EW systems.[7,23] It uses nested list frame representations per LRU, similar to failure-mode effect analysis. The inference is by truth maintenance, with propagated constraints, and a set of domain independent diagnostic metarules. The detection/test selection/failure mode recognition is by a subsequent domain dependent pattern recognition procedure.[21]

IN-ATE (ARC) — A model-basic probabilistic rule-based KBS for electronics troubleshooting has been written,[7] which produces automatically a binary pass/fail decision tree of testpoints to be checked. The search is by the gamma miniaverage tree search.[21] This KBS does not use explicit symbolic knowledge.

ARBY/NDS — This system (used by ITT) is using a LISP based forward-and-backward logic inference scheme (DUCK) for avionics or communications networks troubleshooting.[7] The hypothesis refinement algorithm can be quite cumbersome, and all failure modes must be known in advance.

ACE — This is a KBS for preventive maintenance of telephone cables, by selecting equipments for said maintenance.[7] The knowledge base is a database containing repair activity records. ACE is in regular use.

LES — This is a production rule based KBS for electronic maintenance,[7,31] the various LRUs are described by frames, to which about 50 rules are applied; it is similar to EMYCIN.

SMART BIT — This is a KBS to incorporate false alarm filters and fault recording in built-in-test systems.

STAMP — This is an avionics box failure detection KBS, with test sequences organized by failure history, and dynamic modification of the fault tree.

IDT[10] — This system (used by DEC) uses CAD data to speed up diagnosis.

CRIB[11] — This is a computer diagnosis (ICL) (Brunel University).

APEX 3[38] — This is a shell written in a POPLOG environment; it uses truth maintenance and mixed chaining. If there is a sufficiently likely goal, backward chain from it; if not, forward chain.

RAFFLES — This system diagnoses computer faults at ICL.

RECONSIDER, — This is a diagnostic KBS construction tool (University of California).

SPEAR — This system analyzes computer faults in operations at DEC.

FAULT DIAGNOSIS OF A POWER SYSTEM[41] — This reports about power system diagnosis from rule-based logical fault propagation across relays and circuit breakers alike in inductive inference.

MIND[42,44] — This reports about tester diagnostics with hierarchical KB representation.

Our review shows that most earlier KBS are simple, solve only one type of problem, have a modest size knowledge base, have a rather simple uncertainty handling capability, and used rules as the primary means of knowledge representation. It is also seen that some of the early researchers now prefer the distributed problem solving approach over the production systems approach. KBS are now focusing on using knowledge other that heuristics. Using metarules[4,23] makes these systems more robust decision aids.

III. KNOWLEDGE BASE DESIGN AND KNOWLEDGE REPRESENTATION

The fault detection/test/maintenance knowledge base will consist of several knowledge bases (KB) each dedicated to an independent source of knowledge such as:

Signals	LRU and system structure (layout, causality structure)
Images	Historic maintenance and performance data
Observation reports	Experimental knowledge of maintenance staff
FMEA analysis	
Action lists	Time, location
Maintenance manual	

Each of these specialized KBs appears as a node in the global KB (Figure 1).

Knowledge representation refers to the task of modeling real world knowledge in terms of computer data structures. The basic task is to identify knowledge primitives and to define a system to combine the primitives to represent higher level knowledge.

The problem of representation is closely related to the problem of how the knowledge is used. In KBS, knowledge is used to solve problems, i.e., determine new facts based upon what is already known. The knowledge should be efficiently usable and easily expandable. Knowledge, therefore, has to be represented for quick and easy retrieval, for adequately expressing the various similarities and distinctions, for high computational efficiency, for ease in further expansion and modification, for use in reasoning or solving a specific problem, and for efficient storage.

In order to satisfy the various requirements for the knowledge representation, different techniques have to be used for different types of knowledge.[28] Selecting appropriate knowledge representation schemes is important because AI research has shown that the problem solving strategy heavily depends upon the representation.[20] Selecting an appropriate representation has the impact of either rendering the system successful or a failure. The knowledge representation scheme can effectively limit what the system can perceive, know, or understand.[6]

The basic knowledge representation techniques are (1) symbols, (2) fact lists, (3) semantic networks, (4) logic predicates, (5) rules or production systems, (6) frames, and (7) knowledge objects. These basic representation entities are used to construct representation schemes with desired characteristics for the particular applications. To provide powerful expressiveness,

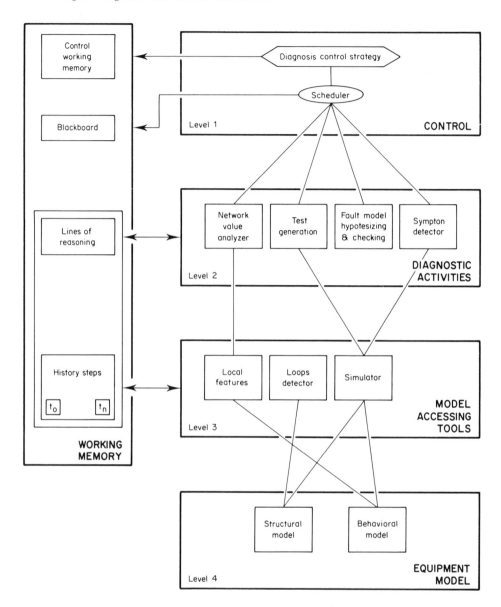

FIGURE 1.   Layered knowledge base for failure/detection/test/maintenance.

conciseness, and easy modifiability by domain experts, the representation schemes evolve into stand-alone, high-level programming languages.

Representation schemes are evaluated on the basis of:

1.   Expressiveness. Does the representation scheme make all of the important distinctions between the concepts being represented?[6]
2.   Structural representation. Can it support different specialized forms of representation?[12]
3.   Computational efficiency. Does the representation scheme allow for efficient computation of various inferences required for the task?[6]
4.   Modifiability. Is the representation scheme easily modifiable?
5.   Conciseness. Is the representation scheme compact, clear, and at the right level of abstraction?[6]

Features	Frames	Semantic Networks	Predicate Calculus	Production System
Expressiveness	10	10	2	2
Structural Representation	8	10	2	2
Computational Efficiency	10	5	3	3
Modifiability	10	1	1	1
Conciseness	10	2	2	2
Representational Uniformity	8	5	10	10
Logical Consistency	5	5	10	5
Easy Retrieval and Access	10	5	5	5
Multiple Level of Representation	10	10	4	4

FIGURE 2.   Comparison of knowledge representation schemes.

6.   Representation uniformity. Can different types of knowledge be expressed with the same general knowledge representation scheme?[12]
7.   Logical consistency. Are the different knowledge units mutually and logically consistent?[12]
8.   Easy retrieval and access. Is the representation scheme such that the desired knowledge can be easily accessed?[12]
9.   Multiple level representation of knowledge. Does the representation scheme allow the representation of the same concept at different levels of abstraction?[4,12]

Figure 2 presents a comparison of four representation techniques or schemes based upon a discussion in Reference 12. Because of its flexibility in representing different types of real world knowledge, and because very little commitment is required to any one representation technique, our preferred choice is a frame based representation scheme.

For large knowledge bases that need to evolve with time and may require frequent updating, both production systems and frame type representations are adequate. Requirements for computational efficiency, expressiveness, conciseness, and easy software maintenance dictate frames over production systems.

The representation schemes will be identified based upon the detailed analysis of the desired characteristics of various knowledge sources such as situation databases, historical database, maintenance manuals, expert maintenance technician experience, and the functional knowledge pertinent to maintenance task.

Frame type representations can be extended to inheritance mechanisms for sharing information to achieve conciseness; another inheritance scheme used[23,25] is the LRU-Block-Module-Subsystem hierarchy, applicable both to structural descriptors, observations, and interfaces between objects.

Failure detection/test/maintenance KBS are required to be able to interface with maintenance personnel with different levels of skill. This requires representing the knowledge at various levels of abstraction. It involves designing mapping mechanisms from primitives to higher levels of representation. Development tools are available that provide this type of layered KB capability.

IV. DISTRIBUTED REASONING TECHNIQUES

The failure detection/testing/maintenance KBS is implemented on a processor to assist the reasoning process of maintenance personnel or test equipment. For each specialist knowledge base node, a specialist reasoning process is designed; this process will also use general inference procedures inherited from the general inference engine of the entire KBS.

A. Communication Between KB Nodes

Communication problems arise because of unrestricted cross talk between the nodes. The proposed approach to this problem is to build in hierarchical communication between the problem solving specialist nodes.[17] Since most of the specialist nodes have to access intelligent database nodes, this cross communication is handled through a blackboard mechanism.[29] In the blackboard mechanism, all KB nodes read data from a blackboard and write their results on the blackboard. Nodes not in direct hierarchical link exchange information through the blackboard without being explicitly aware of each other. This approach mixes the power of direct hierarchical communication whenever possible, and avoids the drawback of cross talk through the use of blackboards.

B. Control Structure

KBS need to have a highly efficient control structure. The purpose of a control structure is to determine which subtask should be done next, i.e., what KB node to access next. In simpler KBS, a common approach is to use production systems type of control, where a list of condition, match, and act rules is evaluated repeatedly. For large KBS consisting of several KB nodes, the issues of control and communication are interrelated. Where possible, a node hierarchy is constructed, and the hierarchy inherently determines the next KB node to be accessed. When not possible, blackboard communication is used. In this case the KB nodes invariably interact only with the blackboard.

Apart from having a control structure at the KB node level, there is also a need for an overall global control. Global control determines when to start the detection/test/diagnosis process, and when the process is complete.

C. Inference Engine

The inference engine uses knowledge in the knowledge base to solve a specific problem by emulating the reasoning process of a human expert. The AI approach to solving a problem essentially consists of searching a solution from a search space. In AI terminology, the set of all possible solutions is known as the search space. The inference engine essentially consists of problem solving strategies that use knowledge in the knowledge base to search for a solution.

The problem solved by the failure detection/test/maintenance KBS is to determine the correct maintenance actions based on the symptoms, repair history, and maintenance.[21] The KBS will consist of several search spaces: the set of possible symptoms, the set of all component malfunctions, the set of all possible tests that can be conducted, and the set of all possible repair actions. The historical database is also a kind of search space.

The search space can be finite and small, it can be finite but quite large, or even practically infinite. The search space can be structured or unstructured. For structured search spaces, the search strategy and the criteria for selecting a solution are both algorithmic. In unstructured search spaces, the search strategy and the criteria for selecting a solution are both algorithmic. In unstructured search spaces, the search strategy cannot be prespecified in an algorithmic way. Expert knowledge in terms of heuristics and trial and error techniques are generally used to search through such spaces. Additional problems arise when the search space is large. Not only is there a problem of time to evaluate each possible solution, but also there is a problem of focusing the problem solving strategy. That is, which nodes should be evaluated and in which order. Human experts are known to successfully perform under these conditions. In expert systems search and focusing problems are solved by the inference engine.

The experience knowledge used by an expert to solve a problem is rarely precise. It generally consists of less-than-certain facts, heuristics from other domains, assumptions which are made unless a contradiction is reached, solutions which are proposed and tested, and some rules and facts which are not used because of constraints or relevancy considerations. To handle various kinds of search spaces and to use imprecise knowledge to search for a solution, AI research has developed many approaches. Some of the important inference mechanisms are:

Heuristic search — In many domains and in particular, in failure detection and maintenance, some personnel can diagnose problems more quickly and effectively than others. When this ability is due to special, domain specific knowledge, acquired as a result of experience, then that knowledge is known as heuristics or rules of thumb. This is the type of knowledge that gives expert systems their power because it permits rapid search; e.g., in repairing an auto, if the car won't start, the battery is examined for failure before the starter is examined. If the battery is faulty, the starter is never examined, unless there are multiple problems.

Generate/test — In this case, the search space is not built a priori, but is built as needed. In effect, possible solutions are generated as the system procedes, and are evaluated shortly thereafter.

Forward chaining/backward chaining — In some problems, it is desirable to determine the hypothesis supported by the given data. Such problems can be solved using either forward or backward chaining. In forward chaining, the reasoning procedes from data or symptoms to hypotheses, i.e., given data, preconditions for the truth of certain hypotheses are tested. This process is similar to pruning a decision tree. In backward chaining, the reasoning procedes from hypothesis to data, i.e., the inference engine first selects a hypothesis to be tested and then seeks for data required to test the hypothesis. If a certain hypothesis turns out to be false, the system can undo all conclusions that preceded or followed the false hypothesis.

Recognize/act — In some problems, the occurrence of certain data in terms of features and symptoms necessitate certain actions. The actions are specified by "IF (features pattern) THEN (take action)" types of rules. The most thoroughly researched and validated approach is to determine/recognize the pattern of features by discriminant analysis, nearest neighbor rules, or syntax driven statistical classifiers.[24,21,23,25] For symbolic feature information only, the inference mechanism then matches the IF part of the rule against the available data, and takes action specified by the THEN part of the rule.

Constraint directed — This approach is usually used in design because it is based on the existence of predefined plans, or plans which can be generically defined. Usually, a skeleton of the overall plan is known and subportions of the plan are specialized or completed in a predefined manner, if they satisfy the certain constraints.

Metarules[4,43] — These are in this instance basic blocks of IF-THEN rules, or hypothesis refinement procedures which rely on exhaustive mapping of causality relations and basic actions in the three dimensional space of physical/functional layout, observed features, or failure propagation/search strategy[22,23] (see Figure 3). They allow for KBS to be developed that can be applied to many projects.

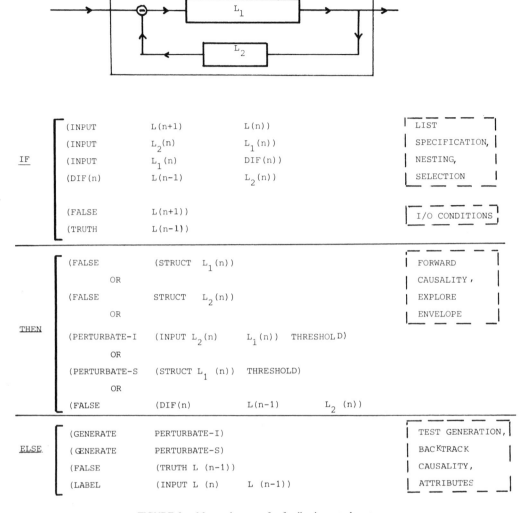

FIGURE 3.   Meta-rule: test of a feedback control system.

The purpose of the global control structure is to coordinate and control the interactions between the node KBS and to provide a friendly interface with the user.[34] The global control performs, e.g., the preliminary analysis of the conditions of the aircraft reported in pilot squawks and determines if the maintenance is needed. It takes into account the time constraints on turn-around time and determines how the problem solving should proceed. The role of the global control system can be compared to that of a maintenance supervisor. The design of the global control is based upon principles similar to the inference engine design.

V. KBS ARCHITECTURE

A. Requirements
Standard requirements put on failure detection/testing/maintenance KBS are[24]

1.   Minimum nondetection probability.
2.   Minimum false alarm probability.

3.   Minimum detection/test selection/maintenance action selection time, especially in interactive usage mode

4.   Knowledge Integration: to enable effective use of procedural heuristic knowledge, specific historical data, fleet data, and a deep knowledge of the functional understanding of the system (includes CAE data)[21.]

5.   Types of Problems Solved: most KBS solve a single type of problem in a narrowly defined area of expertise. The KBS will be required to solve several types of problems in order to perform different tasks.

6.   Resource and Constraint conformance: the KBS should be capable of exploiting inherent redundancy in the system and should take into account the limitations of a particular maintenance squadron in fault isolation and in recommending repair actions. Some of the constraints are time available for repair, personnel/test equipment availability, skill level, and available spare parts.

7.   Expandability and Maintenance: the knowledge base should be easily expandable to accommodate the addition of new knowledge or changes in the current knowledge. The requirement of easy maintenance is critical for large systems.

8.   Capable of Using Meta-Knowledge: in order to handle unforseen situations, the KBS should possess detailed knowledge on how the system works, on general failure modes propagation on what happens in case of LRU failures, and what will happen if certain repair actions are taken. Simulation models are to be included in this class of knowledge.[21]

9.   Capable of Handling Uncertainties: in knowledge base systems, uncertainty arises from three sources — lack of complete data, incomplete pertinent knowledge, and uncertainty inherent in the process. The KBS will have to resolve such uncertainty using redundant knowledge from other knowledge sources or through statistical means.

10.   Explanation Generation and Query Capability: for the KBS to be usable by the maintenance technicians with different levels of skill, it is essential to have sophisticated explanation capability. It must be capable of explaining its line of reasoning and justify the advice it gives to the user. It should also allow the user to query its knowledge base in order to debug or to gain a better understanding about the maintenance task.

B. KBS Architectures

There are three basic KBS architectures: production systems, structured production systems, and distributed reasoning.[28] These architectures must be specified, both for the total KBS, and the specialized node KBS.

1. Production Systems

Production systems (Figure 4) use a single inference engine and a single knowledge base. The inference engine consists of problem independent control strategies which determine which rules should be executed next and executes the actions specified by the rules. The knowledge base consists of a list of production rules IF (features) THEN (actions). All the rules in the list are checked repeatedly until the designed result is achieved. If the number of rules is small then production systems are ideal. However, for even moderately sized knowledge bases, the execution time is very large. For applications that need to have non-delay interactive execution or almost real time execution, this aspect is a major drawback.

2. Structured Production Systems

Structured production systems (Figure 5) divide the knowledge base into knowledge chunks and use metarules to determine which knowledge chunk should be accessed next. Since only a selected number of rules are checked for a given set of data, structured production systems are usually capable of near real-time execution.

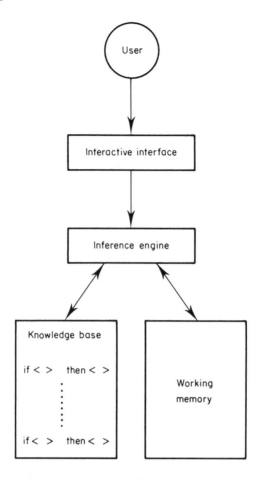

FIGURE 4.   A production system structure.

3. Distributed Reasoning

Distributed reasoning (Figure 6) is based on the notion of specialized KB cooperatively solving a specific inference. Each specialized KBS may be a production or structured production system; the corresponding KB are organized in a hierarchical structure. A blackboard allows all specialist nodes with a structured common area to access information.

C. Comparison

For potentially complex detection/test generation/maintenance systems, distributed reasoning has several advantages:[22]

1.   For a given problem of specified KB size, the execution time is the least[19]
2.   Specialized KB limit the search required, and render execution times relatively independent of KB size
3.   Easy updates

In order to handle unforseen failures, the KBS has to possess some functional knowledge about the system. This requires representation schemes different from the simple IF-THEN constructs. Although in principle any type of knowledge can be represented as IF-THEN rules, it is more efficient to represent knowledge in its natural form. This is important because knowledge and how it is used are two different things. Commitment to an IF-THEN

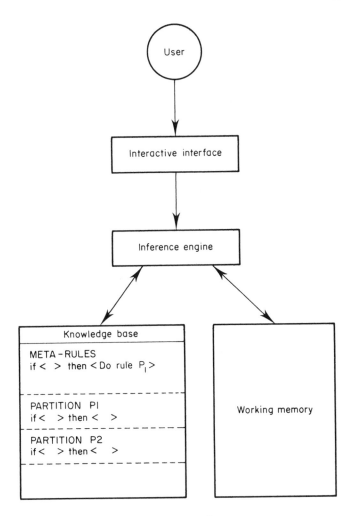

FIGURE 5. A structured production system structure.

structure forces one type of use. For example, knowledge about facts (x is the name of an aircraft), knowledge about system structure (system consists of actuators, flight computers, and sensors), knowledge about processes (to repair electromechanical servo valves, follow step 3.4.2 in the TO number 34), knowledge about events (during a sudden takeoff the computers failed), knowledge about causal relationships (failure of computer will cause following malfunctions), knowledge about goals (the purpose of the mission is to monitor area Z), knowledge about time, and knowledge about actions, do not easily fit a simple IF-THEN type representation.

Production and structured production systems do not allow the use of more sophisticated knowledge representation techniques. In distributed reasoning, each specialist node can have its own unique knowledge representation which is better suited for its particular problem. This feature enables incorporating metarules and functional knowledge into the KBS. A similar problem arises in trying to use situational databases and historical databases in the KBS. Each of these databases will have different structures for these databases, a feature only provided by the distributed reasoning approach.

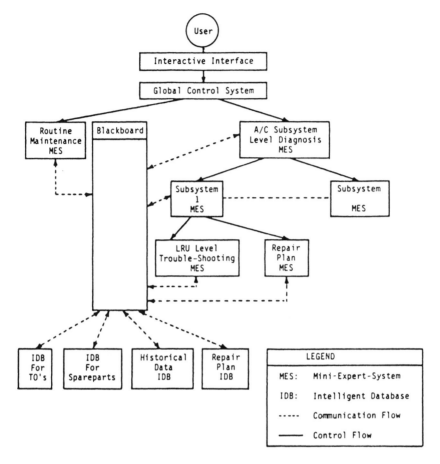

FIGURE 6.   Distributed problem solving structure.

VI. POTENTIAL APPLICATIONS

Five major improvement areas can be considered to have a potential for application, with the use of knowledge engineering techniques:[7,8,2,3,5]

1. Self improving diagnostics: functional test sequences can be cost effectively improved, and automated learning through metarules is a promising area[21]
2. More effective fault detection and isolation, thru built-in-test KBS[22,24]
3. Discrimination between false alarms and intermittent faults, or multiple faults[32]
4. Reduction of skills required for test and maintenance
5. Integrated diagnostics[21]

These improvements are considered likely to go into seven generic failure detection/testing/ maintenance systems:[8]

1. Computer aided preliminary design for testability: testability KBS available during preliminary design phases[1,9]
2. Smart built-in-test (Smart BIT) to identify intermittent faults and reduce false alarms, and carry out recalibration
3. Smart system integrated test (Smart SiT) for system level testing while in operations
4. Box maintenance KBS to provide offline test management with self improvement of functional tests

5. System maintenance KBS
6. Automatic-test program generation (ATPG)[2,23,30,37]
7. Smart bench, which is a maintenance KBS developed for use with bench test equipment controlled by an engineering workstation

Finally, it should be stressed that sensor fusion, which consists in applying distributed reasoning to different information sources (e.g., analog/digital signals, test outcomes, text from verbal reports, inspection images), may lead to novel microelectronic sensors implementing in hardware some of the corresponding knowledge representation schemes.[25]

APPENDIX: TERMINOLOGY

Attribute: a feature or property of an object, usually in frame-based systems.

Backtracking: a search procedure that guesses at a solution and returns to a previous branch point if the guess is wrong.

Backward chaining: a control procedure that attempts to reason back from a goal toward preconditions. Corresponds to a top-down approach.

Blackboard: a data base for recording intermediate results and subgoals during problem solving, often partitioned by level of abstraction.

Expert system: a computer system that uses a knowledge base and an inference engine to find solutions in particular areas of expertise.

Forward chaining: a control procedure that works from preconditions (or subgoals) toward the main goal by applying additional rules. Corresponds to a bottom-up approach.

Frame: a scheme for representing knowledge that associates attributes of features with a given object, employing slots. Attributes can have specific slot values.

Heuristic (rule): an indication or clue as to how to carry out a task — usually contrasted with algorithmic or deterministic approaches.

Inference: a reasoning step or hypothesis based on current knowledge; a deduction.

Instantiation: a specific example of a general class; the procedure of associating specific data objects with a rule or process.

Knowledge base: a data base that contains rules and facts from which an inference engine can deduce solutions.

Knowledge engineer: a specialist who builds a knowledge base by formalizing information gained from experts like physicians or engineers.

Metarule: a rule that dictates how other domain or production rules must be employed.

Rule: knowledge formulated as an "if . . . then" sentence. When the given conditions are fulfilled, it performs the designated actions.

Schema: see frame.

Slot: a feature of a data object or the ability to accept such a feature or attribute.

REFERENCES

1. **Horstman, P. W.,** Design for testability using logic programming, Proc. IEEE Int. Test Conf., Philadelphia, Pa., 1983, 706.
2. **Kunert, A. J.,** ATE-applications of AI, Proc IEEE Autotestcon, Minneapolis, Minn., 1982, 153.
3. **Freund, T. G.,** Applying knowledge engineering to TPS development, Proc. IEEE Autotestcon, Denver, Colo., 1983, 318-321.
4. **Barr, A.,** Metaknowledge and cognition, Proc 6th IJCAI, Morgan & Kaufman, Palo Alto, Calif., August 1979, 31.
5. **Sumner, G. C.,** Knowledge based systems maintenance applications, Proc IEEE Autotestcon, 1982, 472.
6. **Woods, W. A.,** What's important about knowledge representation, *IEEE Computer,* 16, 22, 1983.
7. Proc-Artificial intelligence in maintenance, AFHL, USAF Systems Command, Technical Report AFHRL-RE-84-25, 1984.
8. Artificial intelligence applications to testability, RADC, USAF Systems Command, Technical Report RADC-TR-84-302, 1984.
9. Computer aided testability and design analysis, Technical Report USAF, RADC-TR-83-257, 1983.
10. **Shubin, H., et al.,** IDT: an intelligent diagnostic tool, *Proc. AAAI,* Morgan & Kaufman, 1982, 290.
11. **Hartley, R. T.,** CRIB: computer fault finding through knowledge engineering, *IEEE Computer,* 17, 76, 1984.
12. **Bobrow, D. G.,** Panel discussion on AI, *Proc. IJCAI,* Morgan & Kaufman, Palo Alto, Calif., 1977.
13. **Chandrasekaran, B., Mittal, S., and Smith, J.** Reasoning with uncertain knowledge: the MDX approach, Proc. 1st Conf. American Medical informatics Assoc., 1982.
14. **Davis, R., and Shrobe, H.,** Representing structure and behaviour of digital hardware, *IEEE Computer,* 16, 75, 1983.
15. **Fox, M. S., and Strohm, G.,** Job-shop scheduling: an investigation in constraint directed reasoning, *Proc. AAAI,* Morgan & Kaufman, Palo Alto, Calif., August 1982, 178.
16. **Genesereth, M. R.,** Diagnosis using hierarchical design methods, *Proc. AAAI,* Morgan & Kaufman, Palo Alto, Calif., 1982, 178.
17. **Gomez, F.,** Knowledge organization and distribution for diagnosis, *IEEE Trans.,* Vol. SMC, 1979.
18. **Hollan, J. D., et al.,** STEAMER: an interactive inspectable simulation-based training system, *AI Magazine,* 1984.
19. **Niwa, K., et al.,** An experimental comparison of knowledge representation schemes, *AI Magazine,* 5(2), 29, 1984.
20. **Newell, A. and Simon, H.,** *Human Problem Solving,* Prentice Hall, Englewood Cliffs, N.J., 1972.
21. **Pau, L. F.,** *Failure Diagnosis and Performance Monitoring,* Marcel Dekker, New York, 1981.
22. **Pau, L. F.,** Failure diagnosis systems, *Acta IMEKO,* North Holland, Amsterdam, 1982.
23. **Pau, L. F.,** Failure diagnosis by an expert system and pattern classification, *Pattern Recognition Lett.,* 2, 419, 1984.
24. **Pau, L. F.,** Applications of Pattern recognition to failure analysis and diagnosis, in *Human Detection and Diagnosis of System Failures,* Rasmussen, S. and Rouse, W. B., Eds., Plenum Press, New York, 1981, 429.
25. **Pau, L. F.,** Integrated testing and algorithms for visual inspection of integrated circuits, *IEEE Trans.,* PAMI-5(6), 602, 1983
26. **Pau, L. F.,** An adpative signal classification procedure: application to aircraft engine monitoring, *Pattern Recognition,* 9, 121, 1977.
27. **Clancey, W. J., et al.,** NEOMYCIN, *Proc. IJCAI,* Morgan & Kaufman, Palo Alto, Calif., 1982.
28. **Winston, P. H.,** *Artificial Intelligence,* Addison-Wesley, Reading, Mass., 1983.
29. **Earman, L. D., et al.,** Hearsay-II, Computing surveys, 12(2), 1980.
30. **Kramer, G. A.,** Employing massive parallelism in digital ATPG algorithms, Proc. IEEE Int. Test Conf., Philadelphia, Pa., 1983, 18.
31. **Perkins, W. A., and Laffey, T. J.,** LES: a general expert system and its applications, *SPIE Proc.,* 485, 46, 1984.
32. **Edgar, G. and Petty, M.,** Location of multiple faults by diagnostic expert systems, *SPIE Proc.,* 485, 39, 1984.
33. **Hamscher, W. and Davis, R.,** Diagnosing circuits with state — an inherently underconstrained problem, Proc. Conf. of American association for artificial intelligence (AAAI), 1984, 142.
34. **Hudlicka, E. and Lesser, V. R.,** Meta-level control through fault detection and diagnosis, Proc. Conf. of American association for artificial intelligence (AAAI), 1984. 153.
35. Digest of papers, IEEE Computer society workshop on "Applications of artificial intelligence to fault tolerance", Puerto Rico, January 12-14, 1983.
36. **Barr, A. and Feigenbaum, E.,** *Handbook of Artificial Intelligence,* William Kaufman, San Francisco, 1982.

37. Expert software system for test vector generation and program execution, Fairchild CAD Technical report, Nov., 1981.
38. **Merry, M.,** Apex 3: an expert system shell for fault diagnosis, *GEC J. Res.,* 1(1), 39, 1983.
39. **Basden, A. and Kelly, B. A.,** DART: An expert system for computer fault analysis, Proc. 7th IJCAI, Morgan & Kaufman, Palo Alto, Calif.,1981, 843.
40. **Phillips, B. et al.,** INKA: the INGLISH knowledge acquisition interface for electronic instrument troubleshooting systems, T.R. CR-85-04, Tektronix, Beaverton, Ore., 1985.
41. **Matsumoko, K., Sakaguchi, T., and Wake, T.,** Fault diagnosis of a power system based on a description of the structure and function of the relay system, *Expert Syst.,* 2(3), 134, 1985.
42. **Mullis, R.,** Expert system for VLSI tester diagnostics, Proc. IEEE Test Conference, Philadelphia, Pa., 1984, 196.
43. **Davis, R.,** Metarules: reasoning about control, *Artif. Intelligence,* 15, 179, 1980.
44. Special issue on artificial intelligence techniques, *IEEE J. Des. test Computers,* 2(4), 1985.

INDEX